"Chilling. . . . Junger propels himself into some of the world's most perilous places and situations. . . . His writing is a steely mixture of first-person narrative and thorough reportage. The former gives his work immediacy and the latter gives it depth. . . . Junger's portrayal of the Afghans . . . puts a poignant human face on a people whose suffering surely is not over."
—*Cleveland Plain Dealer*

"[Junger] brings together previously published essays from the world's front lines in a stunningly rich montage of people and death. . . . In *Fire*, Junger takes us as close as we can get to the front lines to watch real heroes without being able to smell the gunpowder ourselves."
—*The Harvard Crimson*

"Junger [is] a literary treasure."
—*New York Daily News*

"Sebastian Junger has a superb instinct for distilling the confusing and complex. . . . The reporting, from the smoldering edge of wildfires in the West to rotting mass graves in Kosovo, is solid and . . . absorbing."
—*Miami Herald*

"Reportage from some of the most desperate corners of the globe. . . . Junger is a literary sensation. . . . He writes better than just about anyone at the moment about elemental forces."
—*The Scotsman*

"A fearless reporter and understated writer who lets his hard-won facts do the work."
—*Seattle Times*

ALSO BY SEBASTIAN JUNGER

The Perfect Storm

FIRE

SEBASTIAN JUNGER

Perennial

An Imprint of HarperCollins*Publishers*

Grateful acknowledgment is made to the following publications, where many of the pieces
in this book originally appeared, in slightly different form:

Harper's: "Dispatches from a Dead War," August 1999
Men's Journal: "Line of Fire," October 1993; "Blow Up," November 1994; "Escape from
Kashmir," April 1997
National Geographic Adventure: "Colter's Way," Spring 1999; "The Lion in Winter,"
March/April 2001
Outside: "The Whale Hunters," October 1995
Vanity Fair: "Kosovo's Valley of Death," July 1998; "The Forensics of War," October 1999;
"The Terror of Sierra Leone," August 2000; "Massoud's Last Conquest," February 2002

HarperCollins books may be purchased for educational, business, or sales promotional use.
For information please write: Special Markets Department,
HarperCollins Publishers Inc., 10 East 53rd Street, New York, NY 10022.

First Perennial edition published 2002.

Designed by Margaret M. Wagner

Library of Congress Cataloging-in-Publication Data is available.

ISBN 0-06-008861-3

02 03 04 05 06 ❖/RRD 10 9 8 7 6 5 4 3 2 1

This book is dedicated to
Ellis Settle, 1924–1993

Contents

Introduction

In 1989, when I was in my late twenties, I saw a magazine photo of half a dozen forest fire fighters taking a break on the fire line. They wore yellow Nomex shirts and hard hats and had line packs on their backs and were leaning on their tools in a little meadow, watching the forest burn. In front of them was a wall of flame three hundred feet high. There was something about the men in that photo—their awe, their exhaustion, their sense of purpose—that I wanted in my life. I tacked the photo to my wall and lived with it for a whole winter.

It was an uninspiring time in my life. I was living in a grim little apartment in Somerville, Massachusetts, I'd quit waiting tables, and I had vague ideas of making my living as a writer. The only good thing I had going on was an intermittent job—more of an apprenticeship, really—working as a climber for a tree company. I'd met a guy in a bar who showed me an enormous scar across his knee from a chain saw accident, and offered me a job. He said he'd teach me to climb if I worked for him whenever he needed someone. I agreed. I climbed trees over houses, trees over garages, trees over telephone lines. I climbed trees that were twenty feet high and swayed from my weight; I climbed others that were 150 years old and had branches so big that holding them was like hanging from the neck of an elephant. Some of the trees had to be taken down; some just had to be pruned. All of them terrified me. I learned to work without looking down. I learned to work without thinking too directly about what I was doing. I learned just to do something regardless of how I felt about it.

Ultimately, I hoped that the work I was doing might lead to a job fighting wildfire. I knew that chain saws were used on fires—one of the guys in my photo had one over his shoulder—and I thought that maybe if I showed up out West with my saw, I could get onto a crew. With fires that big, it seemed that they might take anyone they could get.

That turned out to be emphatically untrue. Forest fires are as much a job opportunity as a natural calamity, and there is a lot of competition to get onto the crews. I made some calls and was told that I had to work for a couple of years on a secondary crew before I could even apply for a full-time position fighting wildfire and that even the secondary crews were hard to get onto. I also needed a "fire card," which meant that I had to pass a training course, but it admitted only people who were already working in one of the government agencies involved in wildfire. I gave up on the idea of fighting fire and stayed East to continue working in the trees.

As jobs go, climbing was hard to beat. I got in very good shape. I lost my fear of heights. I started making very good money. I would bid on jobs, subcontract out the ground work, and do the climbing myself. The amount I made depended on how fast I worked and how well I priced the job. I made two hundred dollars a day, five hundred a day, a thousand a day. Some days I climbed with such confidence that I almost felt that I didn't need to use a rope; other times I was filled with such clumsy fearfulness that I could hardly get off the ground.

My experience as a climber culminated one clear, cold November day, when the owner of a tree company asked me to give him a price on a very dangerous job. A large tree had split down the middle, and the bulk of the tree was still balanced in a tiny piece of trunk. Working in a tree like that would be risky because it was unstable, and if it came down unexpectedly, the climber would almost certainly be killed. I walked around the property, looked at the tree from various angles, and told him, "Five hundred dollars." He shrugged and agreed. It wasn't worth five hundred dollars to go up into that tree—it wasn't worth any amount—but I saw another way to do it. On either side of the property were two taller trees that were roughly lined up with the one in question. I climbed both of the taller trees, set up a tension line between them, clipped into it, and pulled myself hand over hand until I was directly over the tree that had to come down. I rappelled down into it and began working. If it fell out from under me, I was still safe. I limbed the tree out and then dropped the trunk in sections. It took two hours. At the time it felt like the best thing I'd ever done.

Inevitably I was going to have an accident—almost every climber I knew had—and mine came while I was pruning a small elm in Wellfleet, Massachusetts. I was in a hurry, cutting too quickly, and the next thing I knew, I'd hit the back of my leg with the chain saw and exposed my Achilles tendon. At first the wound didn't hurt much and barely bled, but it shook me up badly; if I'd severed the tendon, I could have had problems my whole life. The accident was sloppy and unfortunate, but it made me realize that I didn't want to be a climber and struggling writer forever. I was thirty years old; I should either tackle a book project or get out of the writing business altogether.

The idea for a book came to me gradually, while I was recovering from my injury. What about a book on dangerous jobs? Logging, commercial fishing, drilling for oil: They all were jobs that society depended on, and they were vastly more dangerous than the sorts of adventure sports that were becoming so fascinating to the public. Six months later—with no magazine assignment and certainly no book contract, but with a huge measure of last-ditch determination—I flew to California, rented a car, and drove to Boise, Idaho. One hot day in late July 1992 I presented myself at the smoke jumper loft adjacent to the Boise airport and explained my intention to write about wildfire. The next day, to my amazement, I was in a government helicopter looking down at the Flicker Creek fire.

The result was a long piece on forest fire fighting that I envisioned as the first chapter in my book on dangerous jobs. Parts of it were published as a magazine article, but the rest sat in a drawer while I went on to tackle other projects. The next topic was commercial fishing and focused on a Gloucester swordfishing boat named the *Andrea Gail* that was lost with all hands during a huge storm in 1991. (That chapter took on a life of its own and eventually became *The Perfect Storm*.) Finally, I wanted to write about war reporters, a topic that had an added appeal because I could always try to do that if my book-writing career didn't work out. In July 1993 I flew to Vienna, Austria, and walked into the Associated Press office and asked if they needed any help in Bosnia. The answer was no. I went anyway. Two weeks later I was in Sarajevo, in the middle of the civil war.

I think it's fair to say that I had absolutely no idea what I was doing. I learned quickly, though. I started doing a little free-lance radio reporting; I wrote a couple of newspaper articles; I watched and tried to emulate the other journalists. Half of them were as inexperienced as I was—for the most part, our credentials simply stemmed from the fact that we were there—but we all had one thing in common: We were absolutely mesmerized by what was going on around us. None of the journalists whom I knew wanted to leave the war, ever; none of them felt that it was anything less than the most important event in their lives.

I still don't fully understand why that may be. What is this fascination that roots fire fighters in their tracks while three-hundred-foot flames twist out of a stand of spruce? Why do journalists—I've done this myself—crawl up to front lines even though there's almost no information of any journalistic value there? It's tempting to draw some dreadful conclusion about the inherent voyeurism of humans, but I think that would be missing the point. People are drawn to those situations out of an utterly amoral sense of awe that has nothing to do with their understanding of the larger tragedy. Awe is one of those human traits, like love or hate or fear, that overpower almost everything else we believe in, at least for a little while. Some people experience awe when they are in the presence of what they understand to be God; others experience it during a hurricane or a rocket attack. In a narrow sense, these situations are all the same: They completely override the concerns of our puny human lives.

I never wrote my book on dangerous jobs; my fascination with those kinds of stories developed into a general passion for foreign reporting. The stories in this book all deal, in one way or another, with people confronting situations that could easily destroy them. I should make it clear that as a journalist I was not in their shoes; I was rarely in serious danger, and I rarely lacked a quick way out. I had my own fears to confront, though. Stepping off an airplane to work in a foreign country is one of the most terrifying feelings I know, not because something bad might happen to me but because I'm convinced that I'm going to fail. You have two weeks to understand a completely alien culture, find a story that no one has heard of, and run it into the ground. It never feels even remotely possible.

But it is. And in the process of doing my work, I have had the honor of witnessing, up close, the awful drama of human events. Like the fire fighters in my old photo, I have found it impossible to turn away.

Annapolis, Maryland, 2001

FIRE

FIRE

1992

Late in the afternoon of July 26, 1989, a dry lightning storm swept through the mountains north of Boise, Idaho, and lit what seemed like the whole world on fire.

A dry lightning storm is a storm where the rain never reaches the ground. It evaporates in midair, trailing down from swollen cumulus clouds in long, graceful strands called virga. The electrical charges from a dry storm do not trail off before they hit the ground, however; they rip into the mountains like artillery. On July 26, 1989, lightning was hitting the upper ridges of the Boise National Forest at the rate of a hundred strikes an hour. Automatic lightning detectors at the Boise Interagency Fire Center were registering, all over the western states, rates up around two thousand an hour. By nightfall 120 fires had caught and held north of Boise, little one-acre blazes that eventually converged into a single unstoppable, unapproachable front known as the Lowman fire.

For the first three days Lowman was simply one among hundreds of fires that were cooking slowly through the parched Idaho forests. Around four o'clock in the afternoon of July 29, however, the flames reached some dead timber in a place called Steep Creek, just east of the town of Lowman, and the fire changed radically. The timber was from a blowdown two years earlier and was so dry that when the flames

touched it, the entire drainage went up. The fire created its own convection winds, making the fire burn hotter and hotter until the fire
behavior spiraled completely out of control. Temperatures at the heart
of the blaze reached two thousand degrees. A column of smoke and
ash rose eight miles up into the atmosphere. Trees were snapped in half
by the force of the convection winds.

The fire rolled across Highway 21 and right through the eastern
edge of town, detonating propane tanks and burning twenty-six buildings to the ground. A pumper crew was trapped at the Haven Lodge,
and they hid behind their truck and finally stumbled out of the blaze
an hour later, safe but nearly blind. The fire had attained a critical
mass and was reinforcing itself with its own heat and flames, a feedback loop known as a fire storm. The only thing people can do, in the
face of such power, is get out of the way and hope the weather
changes.

Which they did, and which it did, but not until a month later,
after forty-six thousand acres of heavy timber had been turned to ash.

I saw the site of the Lowman fire in 1992, three years afterward,
when the ponderosa seedlings were already greening the hillsides. A
roadside plaque said that eight million ponderosa and Douglas fir
would be hand-planted by the mid-1990s. The plaque went on to describe how the land had been treated with enzymes so that water and
microorganisms could penetrate soil that was now seared to the con-
sistency of hard plastic. Thousands of flame-killed trees had been
dropped laterally along the slopes to keep the land from washing away,
and thirty thousand acres had been planted with grass and fast-
growing bitterbrush. In a hundred years, more or less, the area would
again look the way it once had.

I was driving a big, painfully beautiful loop from Ketchum, Idaho,
around the Sawtooth Mountains and down the South Fork of the
Payette River toward Boise. It was late afternoon when I drove
through the Lowman burn, and the quiet darkness of the dead valleys
depressed me. The West was well into one of the worst droughts of the

century, and I was out there to see the wildfires that it was sure to produce. My idea was to go to Boise—where all the fire-fighting resources were coordinated—tell them I was a writer, and hope they let me on a fire.

I pulled off down an old logging road and pitched my tent in a clear-cut. It seemed to get dark very quickly that night, and I cooked spaghetti on my camping stove and went to sleep listening to the weekend traffic die down on Highway 21. The Lowman fire, I'd heard, had burned so hot that Highway 21 had melted. There were places, I'd heard, where fire trucks had left their tread marks as they rushed from Boise to fight the flames.

In 1965 the U.S. government established the Boise Interagency Fire Center to coordinate the three federal agencies—the Bureau of Land Management (BLM), the Forest Service, and what was then known as the Weather Bureau—that were engaged in fighting wildfire in America. The Bureau of Indian Affairs, the National Park Service, and the Fish and Wildlife Service were added later, and the name was ultimately changed to the National Interagency Fire Center. Two years after BIFC was established, the Northern Rockies were hit with a catastrophically bad season that culminated in the Sundance fire in northern Idaho. BIFC managed to deploy thirteen thousand men and thousands of tons of supplies, prompting a study by the Office of Civil Defense, which was trying to figure out how to handle a similar crisis in the event of a nuclear war.

BIFC is located next to the Boise airport, across the interstate, south of town. The lobby is filled with the sort of display that, were you even vaguely inclined toward a job fighting fire, would make you move out west on the spot. There is a smoke jumper mannequin in full jump gear, including a wire face mask for when the jumper goes crashing into the treetops. There is a board with everything—food, medical supplies, tools—a jumper needs for forty-eight hours on a fire.

There are color photos of air tankers dropping retardant and sheets of flame rising from stands of trees. One photo shows a fire in dense forest on the Umpqua National Forest in Oregon. "The total-timber jump spot," the caption reads, "trees in this photo are between 80 and 125 feet tall. Five of six smokejumpers committed to this fire 'hung up' in trees, thus the rope carried in the leg pocket for a 'letdown.' The fire was stopped at a quarter-acre."

A short, powerful man named Ken Franz—one of the Boise smoke jumpers, as it turned out—spotted me loitering in the lobby and came over to investigate. I told him I was interested in wildfire, and he motioned me into a cluttered conference room and sat me down at a long table to tell me the basics. Behind Franz was a map of the western United States that covered most of one wall. There were seven red circles on it around seven towns: McCall, Idaho; Boise, Idaho; Missoula, Montana; Redmond, Oregon; Redding, California; Silver City, New Mexico; and Winthrop, Washington. Franz turned and pointed to them.

"Those are the smoke jumper bases," he said. "We are constantly getting sent from one place to another; you never know where you're going to be the next day, or the next week. They just shift resources around to wherever the hazard is greatest."

That shifting of resources, Franz explained, is what BIFC is designed for. Everything from government-issue paper sleeping bags to food to foam fire retardant to Ken Franz himself is shipped around the country, following fires, following thunderstorms, even following droughts. There are 410 smoke jumpers and perhaps 20,000 active and on-call fire fighters in the United States. Should a smoke jumper's father die, say, or his house burn down, BIFC would know what state, what fire, what division, and what 20-person crew he was on. Should an air tanker go down en route from Denver to Missoula—one of hundreds of flights during a busy day fighting fires out west—BIFC would know what route it was taking and when it was supposed to arrive. The immense task of keeping track of all these things is accomplished at the logistics center on the top floor of the main BIFC building, across the parking lot. Across one wall of the room is a huge map of the country. Cardboard cutouts representing airplanes are moved around on it; cards representing fire crews are switched from "available" to "unavailable" slots. More detailed information is stored on a computer. In late August 1987 lightning started two thousand fires across the West that burned almost a million acres. In ten days 22,500 fire fighters and forty-five tons of supplies were deployed to fight the fires. BIFC accounted for every chain saw, every hard hat, every gallon of retardant.

"Smoke jumpers are considered an initial attack force," Franz went on. "That's a generic term for the first response to a fire. The classic situation would be a lightning-struck tree in a remote area where two guys jump in, fell it, buck it up—put out what amounts to a small campfire. Basically, the whole world's a jump spot; within a mile of any fire you can usually find a very acceptable place to land in. On a big fire you have to start somewhere, so you jump a whole planeload and establish an anchor point, at the tail of the fire. You clear helispots for landing supplies, and you work your way around the sides of the fire."

Smoke jumpers land with eighty pounds' worth of gear, including two parachutes, puncture-proof Kevlar suits, freeze-dried food, fire shelters, and a few personal effects. Following them in cardboard boxes heaved out of the airplane with cargo chutes are chain saws, shovels, ax-hoe hybrids called Pulaskis, sleeping bags, plastic cubitainers of water, and dozens of other things needed on a fire. If there's an injured jumper, a medical emergency pack comes out of the plane. If it's a fast-moving fire, the crew can call for boxes of explosives that can blast an instant fire line in the forest duff. The list of what can be thrown at a fire is endless—and expensive. A more cynical view, popular among many, is that the government puts fires out by throwing money on them until it starts to rain.

Not much of the money, however, goes to the fire crews. A jumper makes about $8.50 an hour. If the fire is uncontrolled, as, since smoke jumpers are initial attack, it almost always is, the crews get another 25 percent hazard pay. If they work overtime—again, almost a sure bet—they get time and a half. The jump itself has been ruled as simply another way of getting to the fire, like a bus or a pickup truck, so jumpers get straight pay when they leave the airplane and time and a quarter when they hit the ground. If they are injured on the jump, however, and don't make it to the fire, the hazard pay does not kick in. From fifteen hundred feet it takes about a minute and a half to reach the ground with a parachute. At $8.50 an hour, that's about 21 cents.

"In a good year you can make almost thirty thousand dollars," said Franz. (As with all fire fighters, a "good" year is a year with a lot of fires; a "good" fire is a fire that isn't brought under control too quickly.) "Under twenty thousand is more typical. That's for six months. The rest of the year we sew."

They sew everything: harnesses, fire line packs, jump bags, even little duffels with the BIFC flame logo on the side. They do it to save the government money, they do it because they're better sewers than most manufacturers, and they do it to keep themselves employed. The only thing they don't sew are the parachutes. Some jumpers are certified to make repairs, but the chutes themselves are bought from a manufacturer. The parachutes the BLM uses cost a thousand dollars apiece and are expected, with upkeep, to last at least ten years. They were of a design invented by a French kite maker in the early 1900s. They are called Quantum Q5 Ram Air parachutes.

"*Ram air* means there are cells that fill with air," said Franz. "They make the canopy so rigid you could walk across it. You could also put a line on it and fly it like a kite; in Alaska, jumpers fly their chutes like kites. You steer with toggles and have a forward speed of twenty miles an hour. It's a very high tech delivery system for a very low tech job; once we hit the ground we're just fire fighters. Afterward we have to pack ten miles or more, to the nearest helispot. Our gear weighs over one hundred pounds, and usually we're not even on trails; it's harder work than fighting fires. It keeps you honest."

Honest means capable of enduring a training regimen that used to weed out 30 percent of the preselected men at the training camp (overwhelmingly men, but not entirely). Rookies are considered the fittest and most perfectly trained because they have endured boot camp most recently: three hours of workouts a day, a jump simulator called the Mutilator, an array of courses and tests that virtually guarantee you'll pull your ripcord after jumping out of the plane. Overwhelmingly, it works, though not always. In 1991 a jumper in Montana was killed because he didn't reach for his ripcord until "ground rush," when it was too late. The entire thing was caught on video because it was a training jump. The consensus was he froze.

"The biggest hazard is probably the fire itself," Franz told me. "Felling burning snags, logs rolling down hillsides. Jumping is usually a relief. It's hot in the airplane, and sometimes you feel sick; then suddenly you're totally focused on what you're doing. It's a little dreamlike."

After our talk Franz took me for a quick tour of the jump loft. He showed me the rigging room where the chutes are packed, and the sewing room, and the weight room. Afterward we returned to the conference table, and he popped a short tape into the VCR. It was quick and unprofessional but highly dramatic. It showed a jump crew working a fire at night, right on the line. At one point a sawyer was cutting down a huge ponderosa, and his saw was halfway through the trunk when flames started pouring out like liquid. The tree was hollowed out by fire, it turned out, and was drawing like a chimney. The sawyer kept cutting; the flames kept spurting; eventually the tree fell.

As I left, I asked Franz—against all hope—if there were any fires around for me to see. He told me I'd just missed a good one. An older couple from Pennsylvania had been towing a car behind their RV, and the car got a flat tire; sparks started a fire front two miles wide. Six thousand acres, a million dollars to put out.

"I suspect the government will try to collect too," he said. "Try the dispatch office; they'll know what's going on."

The dispatch office for the Boise National Forest was a trailer east of the airfield. "I just came from BIFC," I told the young woman behind the desk. "I'm a reporter. Are there any fires?"

I felt a little bad asking the question. She didn't blink. "Seven hundred acres as of midnight last night," she said, spreading a map of Idaho out on the table. "Lightning-started, wind-driven, with three helicopters and twenty-two crews. It's called the Flicker Creek fire. They've called in a type one overhead team."

A type one overhead team is called in only when a fire is really bad or is expected to get really bad. The Flicker Creek fire was in steep terrain with extremely dry fuels and strong winds. Steep slopes help a fire because uphill fuels get preheated; winds help a fire because they make it burn hotter and push it across the land. A seven-hundred-acre fire could jump to seven thousand or even seventy thousand in no time at all.

An hour later I was driving north on Highway 21 in my green fire-retardant Nomex pants and yellow fire shirt. In the back seat were a yellow plastic hard hat and a fiberglass and aluminum fire shelter. The shelter is a pup tent that comes in a small pouch with belt loops. It reflects radiant heat, reducing what would be a 1,000-degree fire to 120 degrees or so. I would be assigned a public relations person when I got to the fire camp, the ranger told me. I would be fed and I would be lodged in a tent if I didn't have one. Tomorrow morning a helicopter would take me into the fire line.

I breezed past some bored Forest Service guards and turned off Highway 21, into the hills.

Flicker Creek is one of hundreds of small creeks that cut through the steep, dry hills of the Boise National Forest. Most of the land is grass and rocks and sagebrush, with heavy stands of ponderosa on the north slopes and in the drainages. Flicker Creek empties into the North Fork of the Boise River, which quickly joins the Middle Fork and continues on to fill the Arrowrock and Lucky Peak reservoirs. The entire West was seven years into one of the worst droughts since the 1870s, so both reservoirs were severely depleted. Arrowrock had been reduced to a muddy brook that you could practically jump across.

After twenty miles of rough driving, the road leveled off along the North Fork of the Boise. There was plenty of water up here—or so it looked—and the river was fast and lined with big, open stands of ponderosa. The fire camp was in a huge meadow called Barber Flats that ran alongside the North Fork of the Boise. Hundreds of bright nylon tents were pitched in the yellowed grass. A helicopter thumped over a ridgeline, trailing a retardant bucket. Water trucks rumbled back and forth, spraying the dust down. Hotshot crews came and went, Indian file, or slept in the shade, or sharpened their tools. Some were black with dirt; others looked as if they'd just arrived. They all had on the same green and yellow Nomex that I wore and big lug-soled boots.

I parked my car between the trucks and water tankers and searched out the information desk. The public affairs people knew I was coming, and I was pointed toward a large, deep-voiced man named Frank Carroll. "You'll need boots if you want to go out on the line," he told me. "You'll need water bottles; you'll need food; you'll need gloves. I'll set you up after dinner. You can pitch your tent any-where you like. People get going around five in the morning; make sure you're at breakfast and ready to go by then."

I thanked him and went off to get my gear set up. All around me, big, lean men and a few women went about their duties. I pitched my tent in tall grass behind a cabin that served as a command post and then wandered over toward the catering tent. Behind it was a full-size truck outfitted as a kitchen. Hotshot crews passed by it in line, taking plates of food from the young woman behind the window and then sitting down at folding tables under a canvas army tent. The woman was pretty and had a sheath knife on her belt. I tried to pretend I belonged there, and she loaded my plate up with steak and carrots and mashed potatoes and salad and two slices of white bread.

I took a seat by myself at a corner table and watched the crews come and go, talking loudly, eating fast. Most of the fire fighters were young white men, sinewy and unshaved. There was a scattering of women among them, but the women were treated—as far as I could tell—no differently from anyone else. I didn't think they would be discriminated against so much as subjected to one form of gallantry or another, but they weren't. Everyone seemed to be too tired and hungry to notice the opposite sex. (This turned out to be emphatically untrue.) Furthermore, beneath the baggy clothes and grimy faces it was hard to tell who was what.

The Indians and Latinos generally had their own crews. It was a reflection of demographics more than anything else: Twenty men from Browning, Montana, are likely to be Blackfeet, not white; twenty men taken off the farm crews in the Snake River valley are likely to be Latino, not Indian. It's the Indian crews, the caterer told me, that can really clean out a food truck. "They'll eat anything that's not nailed down," she said. The convicts eat sweets and spicy foods because they can't get much of that in prison. The white 'shot crews are the most health-conscious: They eat a lot of fruits and vegetables; some won't even touch meat.

That was hard to imagine, because the government food contracts were defined by how much protein—meat, in other words—was provided per person per meal. Everyone gets four ounces of meat at breakfast, seven ounces at lunch, and ten to sixteen ounces at dinner. Everything else—vegetables, grains, fruit—was considered a condiment and didn't figure into the equation. It's a lot better than it used to be, though. Back in the dark ages of fire fighting—before women, before showers, before Nomex—the crews subsisted mostly on ham.

Ham and eggs, ham sandwiches, fried ham. Catering trucks were essentially big meat lockers with ham hanging in them, and maybe
some Wonder bread. Back in those days the hotshots wore T-shirts
that said, "When forests burn, pigs die."

After dinner I sat in on the planning session. It was held under a
ponderosa pine by the dirt parking area. The entire overhead team was
there, identifiable by the fact that they weren't dirty and weren't wearing Nomex. They were trained together and used on the Flicker Creek
fire as interchangeable parts of a network called the Incident
Command System. The system is based on the idea that any person
trained for a certain job—logistics chief, information officer, helibase
manager—can perform that job for any agency, in any situation.
Overhead teams are made up of people from a dozen different government agencies and are pulled in from all over the country. You
might find an incident commander from Georgia and an air operations branch director from Colorado and a safety officer from the
next town over. There are seventeen type one overhead teams in the
country, and they mainly fight fires, but they have also been effective
on other catastrophes: oil spills in Alaska; hurricanes in Florida; earthquakes in Mexico. An overhead team was sent to clean up the *Valdez*
oil spill, for example, and the system worked so smoothly that it was
copied by both the Exxon Corporation and the U.S. military. A fire
camp with an overhead team, in fact, can put two people in the field
for every one person acting as support—a ratio roughly twenty times
as efficient as the military's.

I went to bed when the five hundred fire fighters did, at dark. The
only noise was the continuous rumble of the generators. The planning
session had brought bad news, in a way: The fire was cooperating almost too quickly. A thirty-acre spot fire had started in light fuels on
the south front but had been contained by three crews. Seven type two
crews—less experienced than hotshots and usually used for mopping

up—had cut line all the way down to the river, farther than expected. The winds were dying down, and unless they picked up again, the fire would be contained within days.

A fire camp is never completely still. All night long I was aware of the movement of men. They walked past, packed equipment, coughed, spat. Around four in the morning the sounds were so continuous that I woke up even before my watch alarm went off. It was still dark, and the camp undulated with human forms and occasional headlamps. Crews were packing their line gear, drifting toward the catering tent, clustering around the big stand-up kerosene heaters set up at intervals in the field. It was cold, maybe in the twenties. I crawled out of my sleeping bag and pulled on several sweaters and my boots—I had to wear leather on my feet in the helicopter, for some reason—and hustled toward the lights of the tent.

The person assigned to me was Bill Casey, a type two safety officer from the Boise area. He was a strong, clear-eyed man in his late forties who directed the local Bureau of Land Management district and was also qualified to command a type two overhead team. (A type two team handles smaller fires, but operates the same way.) He had bagged thirty elk in the past thirty-two years of hunting them, he said. His father hunts with him and can still shoulder fifty pounds of elk meat, at age seventy-one. Casey is part Blackfeet; he has dead-straight gray hair and brown eyes and a handsome, open face.

"We're a little overstaffed because the fire didn't do what we expected," he admitted as we sat in his truck at the helibase.

"What did you expect the fire to do?" I asked.

"Well." He chose his words carefully. "The guys don't want to see the forests burn, but on the other hand, they want to have a good, productive summer. They like to have two to three weeks on a fire and then move on to another one. In that sense, it's disappointing to have the fire lay down so fast."

We were waiting for our flight into the fire line, and the heater was going full blast in the truck. In the clearing, men were checking the helicopters and writing up flight manifests. The first few days of a fire are usually disorganized to the point of chaos, Casey said. Then it gets better. Casey took advantage of the wait to tell me more about how fires are fought—not from the field but from the office.

"BIFC is just a logistical center that responds to needs in the field," he said. "Suppose Unit X has a fire. As long as it doesn't escape the initial attack, BIFC is not involved. If it does escape, then a regional coordination center gets involved. If the regional center doesn't have enough resources to fight the fire, then BIFC steps in. We've had a lot of fires already this summer, and this is the first one where BIFC has been involved."

Initial attack, he said, comes in many forms. Smoke jumpers are initial attack. Helirappelers are initial attack. Hotshots can be both initial attack and extended attack. Air attack—retardant drops from planes and helicopters—can also be initial or extended. The idea of initial attack is to hit the fire hard and early so that you avoid the expense of an extended campaign. If the attack crews can't contain it, then an overhead team is assembled and put on the fire. A really big fire will suck in 'shot crews from all over the country. There are a total of sixty nationwide. If that's not enough, or if they're needed elsewhere, then type two crews that are trained in fire but often do other work for the Forest Service and BLM are deployed. Farther down in the barrel are convict crews, laborers from the Snake River valley who collect at the center of town when sirens go off, and pickup crews. Pickup crews usually just consist of people with sturdy shoes and in good enough health to pass the training course. When there are pickup crews on a fire, it's a really bad fire. There were no pickup crews on the Flicker Creek fire.

After about an hour the helicopter was ready. It was a Bell Jet Ranger, rented from a private contractor in Arizona for two thousand dollars an hour. The huge Croman logging ships cost three times that and can sling twenty thousand pounds of retardant in a bucket beneath them. We were read the safety procedures by a helitack crewman and then asked how much we each weighed, with gear. The answers were entered into the flight manifest to calculate how much the helicopter could carry once we were out of the ground effect. Once a helicopter rises about a certain height, the downwash from its rotors no longer has the ground to push against, so it can carry less. If you combine the altitude, the air temperature, and the relative humidity, you have a figure called density altitude that determines how much a helicopter can carry. The retardant buckets, called Bambi buckets or Sims buckets, depending on their manufacturer, could even be cinched down to change the size of the container to compensate for variations in density altitude from day to day.

The crewman went over the instructions again: Keep your head down when you come and go from the helicopter. Never leave a helicopter uphill. Never go toward the rear of the helicopter. Never go anywhere that the pilot can't see you. If you violate any of these rules, the attendant is authorized to use force to get you to comply.

"I'll tackle you," he said.

It was the first flight of the day for this helicopter, so the rotors weren't turning when we boarded. Gloves on, helmets and goggles on, sleeves rolled down. The attendant made sure we had buckled our seat belts, and then he closed the doors and the rotors heaved into motion.

"By the way, why couldn't I wear sneakers in the helicopter?" I shouted to Casey.

"Because if we crash, the nylon melts to your skin," he yelled back.

In front of me, the copilot had a flight helmet that said, "Crashing Sucks." The technical term for crashing is *hard landing*. One pilot said that half the pilots he knew had done it. Some had been killed; most hadn't. Many helicopters these days are designed to autorotate, meaning that they don't whistle straight downward if the engine fails.

They do an awkward imitation of a glide. Most of the Boise National Forest is rocky and steeply angled, though, so it was hard to imagine its turning out well one way or the other.

The Bell Ranger lifted off, and within minutes we had put the camp behind us and I could see smoke coming off a distant ridge. It was not the furious orange blaze I had imagined; it was a 750-acre smudge fire that, if touched by wind or left unattended, could spring to life and go ripping off into the heavy timber north of here. Helicopters were shuttling back and forth from Barber Flats, where they dipped their buckets in a retardant tank. The retardant was rust-colored, and you could see it splotched on the hillsides. Below us, one of the big Cromans was easing down into a steep valley to fill its Bambi bucket in a stream. (Hotshots are fond of saying that fish get scooped up in the bucket and can be found flopping on the hillsides. No one had actually seen this happen, but the story became so wide-spread that the Fish and Wildlife Service hired a helicopter to see if its pilot could catch fish on purpose. They failed.) The Croman, Casey yelled, carries a bucket that is too big to fit in the retardant tank back at Barber Flats, so it has to use water. It will go back and forth all day between the stream and the fire, directed by 'shot crews with radios. The pilots are so accurate that they can hit individual trees.

We pounded in low along a ridge and then rose and banked and circled and came in again. I could see yellow-shirted hotshots below us, shrouded in smoke. The fire was down in the valley. Several crews were downslope, strung out along the creeping line of black, and several more crews had just landed at the helispot on the ridge and were waiting for orders to proceed. The helicopter made one more approach and then settled down uneasily onto a ridge that plunged steeply away on both sides. The 'shot crews crouched down against the rotor wash, and the helispot attendant, dressed in a green flight suit and hard hat and goggles, flattened himself behind a boulder as if someone were shooting at him. When the aircraft had set itself down but with the rotors still thumping, the attendant scurried forward in a crouch and opened my door and took me out by the arm. Then he returned to get Casey. Occasionally people back up into the tail rotor or walk uphill into the main one, so there is always someone at the helipad to act as an escort. I dragged my pack over behind some bushes and knelt with my head turned away. The helicopter clattered up over our heads and then continued on to shuttle more overhead around the fire.

Already the sun was hot, and it was going to be a long day of mop-up. Hotshots consider themselves elite, and a day spent cold-trailing—going through a burned area and putting out every ember—is not their idea of fun. It probably isn't anybody's idea of fun, but the type two crews get stuck doing it because by then the 'shot crews have usually been flown to some new emergency.

Today, however, the fire was not cooperating, as Casey said. Unless the wind picked up, everyone would be cold-trailing. The hotshots untaped the axes and shovels and Pulaskis that had been protected for the helicopter flight, and fueled up the chain saws—big Stihl 44s and 56s that the sawyers carried over their shoulders with a pad strapped to their suspenders—and took last swigs of water. It was hot on the ridgetop, but it was going to be ovenlike at the bottom of the valley, where the fire was burning and the wind couldn't reach. Crew by crew, they reluctantly filed off downhill: the Helena Hotshots in their bright pink hard hats; the Flathead Hotshots; the Chief Mountain Hotshots from northern Montana, Blackfeet to a man. Casey motioned to me, and we started off down the slope, following a fire line that the type two crews had put in the day before.

A fire line doesn't look like much: just a strip of ground scraped down to mineral soil that snakes around the perimeter of a wildfire. There are different kinds of fire line; the one we were on was called direct line, or hot line. It was several feet wide and cut right along the edge of the fire—"one foot in the black," as they say. When the fire front is really flaming, the crews cut indirect line anywhere from a few feet to several hundred yards away. Often the area in between is then burned out with torches, called fusees. A burnout uses up the fuels between the line and the fire front, effectively creating a fire line up to several hundred yards wide. A burnout is different from a backfire, which is set deliberately to eliminate huge swaths of fuel in the path of an advancing fire. Often that is the only way to stop a crown fire that is throwing embers miles ahead of it; often, of course, a backfire becomes a disaster of its own. A burnout can be ordered by a crew boss; a backfire can be decided on only by the incident command team.

All handline—as distinguished from catline, which is cut by bulldozers—is made the same way. First, a line scout goes ahead of the crew and flags the route with red tape, taking advantage of anything—streams, rock outcrops, ridgelines—that won't burn. The sawyers follow after the line scout, cutting everything from knee-high sagebrush to 150-foot trees. Each sawyer is assigned a swamper, who pulls the brush out of the way and throws it well outside the line. After the swampers come the rest of the crew, who use rakes, shovels, Pulaskis, and even their hands to scrape the duff clear and remove every shrub or root that crosses the fire line. Untouched forest or range enters at the beginning of this line, and a fuelless strip emerges out the other end. It is supposed to stop fire, and despite its appearance—narrow and insignificant in the midst of such huge geography—it generally does.

Fire line is built and measured in sixty-six-foot sections called chains, a unit of measurement that dates back to the early days of surveying. There are 80 chains to a mile, and hotshot crews should be able to cut 20 chains—a quarter of a mile—of fire line an hour. If it's an emergency, the crew should be able to continue at that pace all day, all night by headlamp, and even all the next day. The unofficial record is sixty-seven hours, set by a California crew boss who had also gone thirty days without a shower. Technically both are in violation of agency policy.

For the time being the fire line served as a very good trail on the steep hillside, and I followed in Casey's dust down toward the crews below us. Fifteen hundred feet farther down was the river. Casey's job, as safety officer, was to walk around all day watching people work and talking to them about potential problems. Do they have a safety zone to retreat to if the fire blows up? Are people being posted as lookouts on each crew? Is everybody wearing a shirt? Most fires are slow-moving, and fighting them is closer to hoeing a garden than being in combat, but when fires blow up, they move with awesome fury, and if people aren't prepared, they die.

"Seventy-two people were overrun at the Butte fire in '85," said Fred Fuller, part of the overhead team of one sort or another ("We've got layers upon layers of us out here," he admitted). He was a lanky, soft-spoken man who was accompanying Casey and me on our rounds for a while. "They were in their fire shelters for an hour and a half. They lived, though—even the Cat operators who didn't have fire shelters. They had to crawl under their machines. But they lived."

Cat operators suffer some of the highest fatality rates on fires because they are reluctant to leave their machines until it's too late; on the Butte fire they had to be dragged off their machines by other fire fighters. Burnovers are considered catastrophes even if no one gets killed, and the people who survive are given counseling within twenty-four hours. By all accounts they are terrifying experiences; that close, a fire storm is as loud as a jet plane taking off, and many of the people trapped in their shelters don't even have radio communication. All they can do is wait and try not to let the convection winds tear the shelters off their backs.

Bob Root, a young severity crew foreman working downslope from us, put it bluntly: "If someone has to get into a fire shelter, then someone else fucked up. I mean, really." Root has fought fire for seven years, since he was eighteen. He was a student at the Colorado State School of Forestry and also a member of the local sheriff's search and rescue team. His severity crew, hired in advance because of severe fire conditions, was fire-ready around the clock and could be on their way within an hour. Root had straight straw-blond hair and a sunburn and dark Bollé glacier glasses. Realizing that I was as good an excuse for eating lunch as any, he sat down and pulled a brown paper bag out of his line pack.

"Probably the worst thing that's happening now is what's called urban interface—that is, houses in the forest," he said. "When it's a question of saving structures rather than just trees, we're more likely to take risks. Basically, if you're protecting a structure and the fire's coming at you, you don't retreat. You stay put and try to work the fire around either side."

Root ate methodically, clearly in no hurry to resume cold-trailing. His bag lunch, delivered by helicopter, seemed geared toward the taste of grade school boys: bologna sandwiches, candy, cookies, more candy, and a couple of apples. As we talked, one of the Cromans came in low over us and started to ease itself down over the helispot. Dangling below it was a car-size bladder bag of water called a blivet. The pilot, looking down out of his side window, placed the blivet gingerly on the crest of the ridge and then released it from the tether. Root didn't take his eyes off it.

"Sometimes when they punch off, the load goes downhill," he explained. I realized that we were, in fact, directly below one and a half tons of water. He took a last look at it as he ate his sandwich.

"This is a pretty unexciting day, but last night was interesting. We were at the bottom of this slope—you never want to be above a fire—and rocks and trees were burning loose and just rolling past us. You could hear them coming. Someone would yell, 'Rock!' and then there'd be total silence while people tried to spot it and get out of the way. Logs would roll to the bottom and ignite the hillside, and the fire would come right back up at us. The night went by fast."

According to Root, all over the West, fire conditions were about as dangerous as they could get. The drier the fuels, the hotter they burn and the faster the fires spread, and fuel moisture levels that should be at 15 or 20 percent are down in the single numbers. Low relative humidity and unstable air (wind) compound the problem.

Fires are generally slow-moving creatures, moving a few chains an hour. But sometimes they can explode up a hillside or across a canyon, and the mountains all around us, scorched by seven years of drought, were as likely as they'd ever been to produce such behavior.

"I've never seen it so dry," said Root, fingering the yellow cheatgrass around us. "One little ember in this stuff, and it ignites; last night every ember was taking. The fuels in Washington and Oregon are drier than what you'd buy at a lumber store. Now all it takes is one lightning strike in thousand-hour fuels for it to catch. That's almost unheard of."

Thousand-hour fuel is a piece of wood between three and eight inches thick. The thousand hours mean that if the fuel were completely saturated with water, it would take a thousand hours for it to lose 63 percent of its weight through evaporation. Conversely, if it were bone-dry, it would take a thousand hours to soak up 63 percent of its weight in moisture. Sixty-three percent is used as a benchmark because it is midway between two points—above 78 percent and below 58 percent—where moisture absorption or evaporation happens in a predictable, linear fashion. In that middle range, however, wood gains or loses moisture in a very complex way, and 63 percent is at the mathematical center point of that nonlinear range. Grass and twigs dry out or saturate almost immediately—one-hour fuels. Sagebrush and other small growth are considered ten- or hundred-hour fuels. Thousand and ten-thousand-hour fuels include everything else, up to ponderosa pine with six-foot diameters. It is very rare for thousand-hour fuels to be as dry as one-hour fuels, but they are. Everyone was worried. As one hotshot said, "There are no small fires anymore." Everything that ignites tends to explode.

Fuel moisture levels can be established by field tests or by extrapolations based on information from the National Weather Service. The fuel moisture level is factored into an index that includes weather conditions, wind speeds, fuel loads (the total oven-dry weight, per acre, of all the fuels in the area), and drought conditions. The information is processed by the National Fire Rating System, which determines the fire risk for every climactic region in the country. It predicts such things as fire intensity, the likelihood of lightning strikes, and what is called the ignition component, the probability that a single firebrand landing in dry fuels will start a fire that requires a response. In addition, precise and highly accurate "spot" forecasts can be derived by the National Weather Service for areas as small as a quarter acre. A spot forecast may state that the temperature range for a specific canyon in the Boise National Forest will be between seventy-eight and eighty degrees; the humidity between 12 and 14 percent; and the winds ten miles an hour. Such information is crucial when 'shot crews are deployed in high-risk situations.

At the macro end of the spectrum is the general state of drought. In 1988 the whole country suffered through a brutally hot summer: Barges were running aground on the Mississippi River; railroad tracks were warping in New Jersey. The drought culminated in the fires at Yellowstone National Park. From late June until early November fires burned a total of 1.2 million acres across the West, more than half of which was in Yellowstone itself. On August 20 alone—Black Saturday—seventy-mile-per-hour winds fanned three-hundred-foot flames through a total of 165,000 acres. Although much of the country returned to normal that winter, the drought in the western states never really ended. Drought-stressed trees died by the thousands. Dry lightning storms continued to ignite huge fires. Fuel moisture contents dipped to 2 and 3 percent. The Palmer Drought Index—a water accounting system that measures water gained and lost in a given area—indicated that the entire West was entering the driest period since records had been kept in the 1870s. On the Palmer Drought Index, 0 is normal, +4 is extreme flooding, and −4 is extreme drought. During the summer of '88 Yellowstone registered −5.8. During the Flicker Creek fire the National Weather Service declared the Boise area to have hit −6.5.

Out here that is not an abstract figure; it has the ability to stun. I repeated it to a division superintendent on another fire, and he jerked his head back as if he'd been slapped.

Root finished his last candy bar, and we both stood up. I thanked him for the talk. Helicopters thumped continuously overhead with the Bambi buckets, and 'shot crews raised plumes of dust on the steep hillsides. Casey angled slowly uphill toward me in the high August sun, and after he arrived, we contoured across the ridge and headed for the west side, where the Chief Mountain and Flathead 'shot crews were cutting snags and putting out spots above the river. Snags (standing dead trees) are routinely cut down whether they pose a fire threat or not; some are completely hollow inside and standing on a virtual toothpick. When they let go, it is without a sound, and occasionally there is someone standing underneath. At the Red Bench fire in 1988, a hotshot was killed by a snag while waiting for a bus by the side of the road. "Death comes from above," one hotshot warned, explaining why he walked around with his head tilted up.

We stopped to talk for a while to the Flathead crew, two of whom had been at the infamous Dude fire in Arizona when a convict crew were burned over in their shelters. Melissa Wagner, now reclined in the cheatgrass eating lunch, recalled hearing the screams of the convicts over the radio as they died. One of the survivors left his fire shelter too early and emerged from the flames with his hair smoking and 47 percent of his body burned. Six others died. Wagner kept fire fighting because she needed a way to pay for law school.

Casey and I continued toward the Chief Mountain crew, visible on a distant ridgeline with smoke coming up behind them and helicopters circling above. Halfway there, trudging through the dust and baking heat, Casey suddenly jumped back and almost knocked me down. A big rattler was coiled on a rock in the middle of the fire line. It didn't move, and it didn't rattle; in fact, it was headless. We reached the Blackfeet crew twenty minutes later. They were sitting in the shade of a small ponderosa, eating their lunches; way downhill, a solitary tree was torching.

"So, you guys know anything about that rattler back there?" Casey asked.

There was silence. Every man in the crew looked off in a different direction.

"Rattler?" Glen Stillsmoking, the crew boss, finally said.

"Sons of bitches," Casey muttered, shaking his head but with a smile.

We sat down in the shade to drink some water and take in the view. One of the crew pulled an obsidian arrowhead from his pocket; he said he'd found it at the helispot at Barber Flats. Stillsmoking leaned in to take a look. "We were a pretty hostile people," he admitted, looking around at his crew. "We ran Chief Joseph off; we were the last to settle down. But now we make good fire fighters. My father was a fire boss; I wanted to go to flight school, but fire just drew me away. We've been to Alaska, Florida, paid vacations—that's what we call work—everywhere. At the safety briefings in Florida they told us to watch out for the alligators."

Downslope, a pair of sawyers—sawdogs, they're called—were dropping snags; the sound of their Stihls reached us as a puny whine. A lone ponderosa burned lazily behind us a mile away, and the river glinted far below us in the canyon. This was probably as peaceful as it got on a fire line, and the crew didn't seem in a particular hurry to ruin the moment. I leaned back and tried to stop wishing the fire would do something big.

The season blew up a few weeks later. I was back on the East Coast when I got a call from Frank Carroll, the information officer for the Boise National Forest, who said that lightning strikes were now starting fires by the dozen at the upper elevations. A twenty-year-old fire fighter had been killed by a snag at the tiny Cascade fire; it hit her so hard that her hard hat was driven into her head. A smoke jumper broke his pelvis while landing on the Red Mountain fire. The town of Cuprum had to be cleared of all thirty residents when the Windy Ridge fire detonated into a five-thousand-acre blaze in one afternoon. A Diamond Mountain hotshot on the Horsefly fire was knocked five hundred feet down a hillside by a flaming log because he had shoved a friend out of the way before trying to dodge it himself. A total of eleven thousand people were on the fire lines at one time, Carroll said, and water levels were so low that Boise-area farmers had been shut off two months early.

A big runaway fire was almost inevitable, and it finally hit at the end of August. On the afternoon of Wednesday, August 19, a thunderstorm swept past Boise and lightning ignited the rangeland east of town. The fire quickly overwhelmed the BLM engine crews sent to deal with it and made its way up into the Boise foothills, leaping the quarter-mile-wide canyon of the South Fork of the Boise River with ease. By then it was in steep terrain and flashing through the grass and sagebrush almost faster than a person could run. Temperatures were in the nineties, and the relative humidity had bottomed out at 5 percent. Conditions were so fast that the fire encircled the town of Prairie and almost torched it, forcing the evacuation of all one hundred residents. The only store in town—complete with a hitching post and a bar—was selling T-shirts that said: "We interrupt this marriage to bring you the fire season."

It was called the Foothills fire, and I got on it ten days after it first ignited. Snags in the timber were dropping at an estimated rate of forty an hour, prompting overhead to pull crews off the lines at night. This in turn prompted locals to say that the fire fighters weren't working hard enough. I was assigned to the Union Hotshots out of La Grande, Oregon. Not only was Union one-third women, but its crew boss, Kelly Esterbrook, was one of only ten women ever to have made it through the brutal smoke jumper training course. The Union Hotshots were one of three crews guarding a strategic section of the fire line and if temperatures stayed high and the wind picked up, they would be right in the middle of things. The fire could roll right across the handlines and into a big stand of diseased ponderosa in the high country, and if it did that, it would be virtually unstoppable.

A public information officer named Karen Miranda had been

assigned to take me up the fire line to meet the Union Hotshots. I picked up my Nomex clothes and fire shelter at the Forest Service headquarters in Boise, bought notebooks and blank cassette tapes, and raced east toward the town of Prairie. It was thirty miles away across a huge swath of dead black rangeland. I had to make the fire camp before three o'clock, when the helicopters started making their evening runs out to the crews, and I got in with a half hour to spare. When I walked up to Ed Nesselroad, the senior fire information officer, he was telling someone about a live elk that had been found with its eyes burned out. He quickly rounded up Karen Miranda, and by late afternoon we were strapped into the canvas seats of an Evergreen helicopter and waiting to be shipped to helispot six—H-6, as it is called—northeast of Prairie.

We got the same warnings as before, except that this time we were shown the emergency shutoffs in case we crashed and the pilot was unconscious. (Never get out of a downed helicopter with the rotors still turning.) We lifted off and immediately could see a cauldron of smoke boiling up out of a valley. Dozens of helicopters flew in and out dumping retardant, and away to the south stretched an endless carpet of blackened rangeland. Trees torched below us, and the sweet, sharp smell of smoke filled the cabin. The helicopter made several low passes at H-6 and finally settled down after the helispot crew had cleared some cargo out of the way.

H-6 was on a ridge just below a vegetation line where ponderosa gave way to alpine fir. Just over a hill was tiny Smith Creek Lake, cupped in the topography like a jewel in some huge brown palm. Tents and bedrolls were scattered through the ponderosa. Hotshots lounged with books or talked in small groups or just sat and stared. A slingload of gear waited to be sorted below the helispot, and near it was a huge pile of cardboard boxes and plastic buckets. It was our food for two days.

An inflamed red sun was setting over the ridge, and smoke was pumping out of a valley to the west of us. Occasionally a tree torched on the ridge and then died down, lighting the campsite—it was dark by the time we ate—with a dull glow. Hotshots stopped eating and turned to watch. These were people who couldn't even remember how many fires they'd been on, I thought. Still, they couldn't keep from looking at open flame. There are stories of crews getting overrun by fire because they were too mesmerized by it to run away.

The camp we were at was termed a spike camp, and the hotshots here—sixty of them plus several helitack crew members—were said to be spiked out. That meant that they were established in a roadless area and supplied by helicopter with food, tools, and paper sleeping bags (marginally warm but washable and reusable). According to Forest Service policy, hotshots should not be spiked out for more than two days in a row. One level less comfortable than a spike camp is the coyote camp, and hotshots are not universal in their love of coyote camps. Coyoteing, as it is called, means dropping in exhaustion wherever you happen to be when it gets dark. Because hotshots have only their line packs when they fight fire, they are usually caught without food, sleeping bags, extra clothes. If it's cold, they will make a fire in the black—the burned area—and huddle around it all night. If it's really cold, they might decide to keep building line simply to keep warm. For food, they might have thought to pack some military MREs (meals ready to eat). If not, they go hungry.

The trees torched intermittently all night, I would wake up and see their glow. The sound of flames consuming pine trees one by one easily carried the mile to camp. I tried, and failed, to imagine what it would be like to be burned over. In 1910 a fire storm overtook a group of forty-five fire fighters who were trying to escape an Idaho fire called the Big Blowup. The leader, Edward Pulaski—who later manufactured the tool that bears his name—led them into an abandoned mine shaft and had to keep them there at gunpoint because they were so terrified. Five men died; the rest emerged several hours later, burned and dazed.

They had survived a fire storm of such ferocity that entire hillsides of timber had been flattened by fire and convection wind.

The hotshots were up well before dawn, as usual, rustling about quietly with their head lamps on. It was very, very cold at that altitude at that hour, and I put on everything I had and wandered down the hill to the kitchen area. There would be a briefing at six while people ate, and then they would attack the fire. The smell of smoke still permeated the camp, but the fire had quieted down during night, and trees no longer torched along the ridge. Fires generally "lay down" after dark because the temperature drops, meaning that the relative humidity rises. People poured themselves cereal and coffee and sat on the hillside, eating and listening to the morning briefing.

The division supervisor, a gruff, stocky man named Fred Bird, stood on a crate in the half-light and projected the plans for the day out across the mountainside. "Okay, we're gonna try to hold our own on that ridge," he said. "We've got good air support, and they're going to work us here all day today and then bump us back to camp. Tomorrow they'll get in some type two crews to hold the ridge and send 'shot crews up to the North Zone to chase spots down that canyon."

One hundred and seventeen miles of fire line had been built, there were only ten more miles to build, and overhead was saying that the fire was nearly contained. Everyone knew that one good spot into some timber could start the fire running again, though. Bird's briefing was short and to the point, and sunlight was just touching the upper ridges when the three crews shouldered their packs and started off up the hillside. I fell in with the Union Hotshots, followed by Karen Miranda, and we worked our way slowly up a steep drainage that ended at a ridge, heavy timber on the far side. Burned timber. In the absence of wind, the ridge had acted as a fire line, and Union was simply going after the spots that had made it over. Spots were anything from a few square feet of ash to a vigorously flaming snag surrounded by an acre of black, but they all had to be put out. We were in sparse alpine fir now, Kelly Esterbrook said, and fire in alpine fir is very hard to fight. The trees blow sparks as they burn, igniting spot fires everywhere, and they grow in dense groves that provide critically high fuel loads when the fire reaches them. The flames climb up into the trees from the lower branches, which often reach to the ground, and make their way up into the densely packed crowns. A crown fire is particularly hard to stop because the fire moves from tree to tree without ever touching the ground. You can stop a crown fire only by cutting a lot of trees down, and after a certain point it becomes a case of cutting down the forest to save it.

The Union crew spread out in pairs to hit the spot fires and Miranda and I climbed on to the crest of the ridge. Below us, an entire river drainage of charred trees stretched away to the west. Occasionally a solitary tree torched down in the valley and sent up a tremendous plume of smoke. A temperature inversion had trapped the smoke low in the valley, and Miranda said that when it lifted—when the smoke started to rise—that meant that the air was turning over, and convection would invigorate the fire. At that point they would probably call in a substantial air attack in an effort to keep the fire contained on the west side of the ridge.

We followed the ridge, which climbed toward a peak that we had seen above camp. Below us on our right, the forest smoked. Below us on our left, hotshots worked the spot fires in small groups of three or four. An enormous Siller Brothers Skycrane helicopter with a bucket clattered up and down the valley, unleashing two thousand gallons of Smith Creek Lake water at a time. Four sawyers from the Smokey Bear crew dropped a flame-gutted ponderosa and then called in a water drop, and the Skycrane responded in minutes. Two thousand gallons, hitting a hillside from a hundred feet, is practically enough water to body surf downhill through the trees. When the water subsided, the sawyers bent over and began grubbing through the wet soil for embers.

Twenty minutes later, as they were finishing up, a message came over their radio from the fire camp: "Steve Shaeffer, Steve Shaeffer,

Smokey Bear, your wife is in the hospital at this moment having a baby." The sawyers grinned at one another; it was rare that overhead called with personal news that wasn't bad. Shaeffer was farther up the ridge, hitting spots alongside the Negrito crew. He wasn't able to go home, but at least he'd gotten the news.

At the top of the ridge we were met by the branch director, a strapping dark-haired man with a white-flecked beard and wind-burned face named Mike Rieser. He had fought fires since 1973 and was now fire control officer for the BLM Craig District in Colorado. We sat down and opened our lunches on a rock outcrop that over-looked the burning valley.

Rieser has personally known eight people who died on fires. "Wildland fire fighting has one of the highest incidences of fatalities and injury in the country," he said. "In 1990 twenty-three people died, out of ten thousand active fire fighters. Six died on the convict crew at the Dude fire in Arizona, and that same week two were burned over in California. I saw a film of the first walk-through after the Dude fire; the heat varied so much that one shelter would be fine and the next one would have started to disintegrate."

The Dude fire happened along the Mogollon Rim, on the north side of the Grand Canyon. It was a classic plume-dominated fire, but the topography of the rim served to intensify the downdrafts. Moments before the fire blew up, the Prescott Hotshots noticed a strange calm that often precedes a plume-dominated situation and radioed the overhead team that they were pulling out. The Perryville inmate crew and the Navajo Scouts crew were warned of the danger as well, but they were in exactly the wrong spot. The downdraft hit right in front of them and, funneled by the contours of the canyon rim, drove the fire straight toward them. Half the inmate crew es-caped, as did the entire Navajo Scouts crew. The rest of the fire fight-ers dived into their shelters and waited for the flame front to pass over them.

From laboratory tests, researchers know that the adhesive that holds fire shelters together starts to melt at six hundred degrees Fahrenheit. That causes the fiberglass and aluminum layers to delaminate, in turn leading to rips and holes in the shelter. After the Dude fire, investigators found hard hats that had melted, leather gloves that had shrunk down to a couple of inches, and fire shelters that had begun to delaminate. Six men died, all from breathing superheated air. In all six cases they died because they had not deployed their shelters properly or because they had tried to leave them too quickly. Some of the survivors also suffered terrible burns, which suggest another possibility. The men who left the safety of their shelters may have done so because they thought they were dying inside them.

Mike Rieser and I sat on a rock outcrop eating our lunches and watching the fire. The sun was very hot and trees torched occasionally down in the valley. Every five minutes the Skycrane went by with a tremendous whump of rotors and released another two thousand gallons of Spring Creek Lake. Clouds began to move in above us, and Rieser said that they were called lenticular clouds and that the sculpted tops meant that the upper-level winds were over one hundred miles an hour. If those winds made it down to ground level, they would have a catastrophic effect on the fire. It was his job as branch director to watch out for things like lenticularis formations or castellanus or stage three cumulonimbus or anything else that might make the fire blow up. Two days earlier Rieser had pulled two entire divisions out of a canyon because he hadn't liked the way the fire was behaving.

"We might not look like we're doing much, but if you're down in the valley digging, you don't see what's coming at you," he said. "Air can rise off a fire, cool in the upper atmosphere, and rush back down. It superheats as it comes down and can overpower the wind field. That's called a plume-dominated fire. It defies prediction. It's what killed the people at the Dude fire; the 'shot crews recognized plume-dominated conditions, but the convict crew didn't. No one could get word to them in time."

Stage three cumulonimbus, thunderclouds, are a particular hazard. Not only do they introduce more lightning into the situation, but the air beneath their thirty-thousand-foot heads is extremely unstable. They can generate downward-moving winds of as much as one hundred miles an hour that hit the ground and spread out in a tremendous circle. The effect is to intensify the interaction among all three components of what is known as the fire triangle: fuel, oxygen, and heat. Ground winds spread dry, hot air through as yet unburned fuels, resulting in more fire and more heat, which in turn circulate more air, spreading the fire even faster. The result is a feedback loop that can be brought under control only if the relationships within the fire triangle break down—either by lowering the temperature of the fire or by depriving it of fuel. Water and retardant drops can bring the temperature of a fire down, and fire lines can deprive it of fuel. Otherwise, the fire spreads until the weather changes or there's nothing left to burn.

In the Northern Rockies wildfire is usually started by lightning. Any lightning strike that reaches the ground can cause an explosion, but only lightning with a continuous current can start fires. In the Northern Rockies it has been estimated that one lightning stroke in twenty-five is a cloud-to-ground stroke capable of starting a fire. The inital bolt from cloud to ground moves relatively slowly, at one two-thousandths the speed of light, but it returns at one-tenth the speed of light and heats the gases inside it to fifty thousand degrees Fahrenheit. That much energy hitting a tree instantly raises it way past the ignition point, and often the tree just explodes. Flaming chunks of wood are hurled into the forest, and if the conditions are right for fire, the flames take hold.

Fire is a chemical reaction that releases energy in the form of light and heat. In the case of a wood fire, the energy was originally derived from the sun during photosynthesis and stored in the plant as cellulose and lignin. Heat—from a fire that is already burning or from a lightning strike—converts the cellulose and lignin into flammable gases, which are driven out of the wood and combined with oxygen in a process called rapid oxidation. At the base of any flame there is a clear band of superheated gases that have not yet ignited, a thin blue area of ignited gases, and a broad yellow band of incandescent carbon particles. The heat generated by this process continues to drive flammable gases out of more fuels, which burn and generate more heat. The heat also drives out moisture that might impede combustion. As long as there is sufficient air, fuel, and heat to ignite more fuel, the fire will keep advancing. As long as the fire keeps advancing, the fire triangle remains stable and continues to create the conditions necessary for fire.

A plume-dominated fire, more commonly known as a fire storm, is this same cycle writ large. In this case the three legs of the fire triangle not only provide the conditions for fire but amplify one another in an apocalyptic feedback loop—a "synergistic phenomenon of extreme burning characteristics," as it is known. As in all fires, heat generates wind, which makes the fuels burn hotter, generating more wind. If a high fuel load is introduced into this loop, a convection cell of smoke and gases can be set in motion that overrides the local wind patterns. During World War II Allied bombers intentionally started fire storms in the German cities of Hamburg and Dresden; in that case the high fuel load was densely packed houses ignited by thousands of tons of ordinance. Once the convection engine has started, it is nearly impossible to stop. Entire stands of trees torch as one. Tornadoes twist through the interior of the storm. Superheated fuels appear to combust spontaneously in a phenomenon called "area ignition." Such a fire can rip through well over one hundred thousand acres of timber in one day.

Another sort of apocalypse, equally destructive, is the running crown fire. Crown fires occur when flames climb so-called ladder fuels into the treetops and are swept along by high winds. In 1967 a running crown fire crossed the Idaho panhandle on a four-mile front that incinerated sixteen miles of timber in nine hours. The Sundance fire, as it was called, was calculated to have burned at rates of up to 22,500

British thermal units per square foot per second. By comparison, 500 Btu is the outer limit of what humans can control; 1,000 Btu describes potential fire storm conditions. The Sundance fire was estimated to release the energy equivalent of a twenty-kiloton Hiroshima type of bomb exploding every ten minutes.

Not all big fires are fire storms, of course, and not all fire storms are big. The Steep Creek blowup on the Lowman fire had the physical characteristics of a fire storm but was limited in area; the Foothills fire developed several convective columns over heavy timber but went on to become a wind-driven fire that ripped through two hundred thousand acres in two days, making it one of the biggest fire runs ever. In the Northern Rockies there are a host of winds that push fire: jet stream winds that drop down over mountainous areas; chinook winds that plunge downslope because of an air pressure differential; cold fronts that move in for twenty-four hours at a time; and, of course, unstable air associated with the fire itself. Unstable air rises and falls with the atmospheric conditions, whistling up canyons, sheering over ridges, bending around solitary trees or boulders to start whirls up to four thousand feet high. Any of those winds, in the wrong situation, could cause a blowup and kill people. That was why Mike Rieser was on a ridgetop watching the clouds rather than down in some canyon fighting the fire.

Hotshots have been known to complain that overhead—the men and women who risk other people's lives—do not do enough. Not only that, but hotshots believe that many overhead have never really fought fire and therefore can't be trusted to make life-and-death decisions. Sometimes that is true; there are the inevitable instances of 'shot crews simply saying, "No, we won't go in that canyon," or, "No, we won't try to hold this ridge." More often, however, the members of a command team, like Rieser, have worked their way up from grunthood to positions of authority over the course of years, if not decades. Rieser has fought fire for nearly twenty years and had two extremely close calls (that he told me about). Once he and his crew fell asleep after cutting line all night and were almost burned over; another time he was caught in a chaparral fire outside Los Angeles. Chaparral fires are extremely volatile because, invariably, the fuels are bone dry, the terrain is steep, and the winds are terrible: Santa Anas that hit seventy miles an hour for days on end. It was in 1979, and Rieser was on a type two crew that, he says, had violated just about every watch out rule in the book.

"We were backfiring off a road two-thirds the way up a ridge," he said. "We couldn't get a good burn because of a marine air intrusion. We double-shifted into the next day and got a dominating Santa Ana wind, and the fire just blew up. We were right in a canyon, it was acting like a natural chimney, and the flame front was on us in about ten minutes."

In those ten minutes the crew managed to jump onto a tanker truck and make it to a marginally safe area at the intersection of two dirt roads, not big enough to qualify as a legitimate safe area but better than nothing. They parked the truck and crouched down between it and a road cut. The crew were so rattled that they were reading the Spanish side of their fire shelter instructions, not understanding a word. While the rest of the crew were trying to figure out their instructions, the fire went through.

"It was so loud that we couldn't even shout to each other," Rieser said. "It was not intolerably hot; the smoke was what was hard. We called in an air tanker and heard it make the drop about half a mile away; that was scary because we realized the smoke was so thick they didn't even know where we were. It was an out-of-control situation, and our fate was in the hands of people who had made very questionable decisions. That was the turning point for me. I've been on worse fires, but I always pull the crews out before it gets bad. They say, 'Aw, we could've held that,' and then they watch it boil over."

———

In 1871 a forest fire swept over the town of Peshtigo, Wisconsin, and killed more than fifteen hundred people. A fire in Chicago killed another three hundred people on the same day and is known as the Great Chicago Fire; the Peshtigo tragedy isn't known as anything. Fires come in waves or complexes, and the Peshtigo and Chicago tragedies were part of a wide swath of fires that year that extended from Ohio to the High Plains. Since 1900 well over seven hundred people—it's not known exactly how many—have died on wildfires in America, the vast majority of them men who were employed or had volunteered to fight the fires. The mass tragedies are mostly from the early days when there were no radios, no fire shelters, no aircraft, and no accurate weather forecasts. The next big fire complex after Peshtigo was the Big Blowup of 1910. Eighty-five men died battling the fires, some of them because they panicked and committed suicide after the fire lines were overrun. State troopers played taps over the caskets and buried them in mass graves in the hills. It was just this side of war.

In the end five million acres were burned in 1910, and flame-killed trees provided fuel for reburn cycles that lasted well into the 1930s. Reburns spread into healthy timber and ultimately redoubled the acreage that was lost. Faced with the possibility of a lumber shortage and consequently under tremendous political pressure, the Forest Service finally gave fire suppression top priority. Money was appropriated by Congress, crews were organized, lookout towers were built, timber companies constructed access roads into the mountains, telephones began to replace runners and mounted messengers. Fire fighting had finally entered, as one historian said, its heroic age.

The new approach showed results, but the West was still vulnerable to large-scale conflagration. The next big round of tactical changes came, not surprisingly, after the next big catastrophe, the Tillamook burn in 1933 that laid waste to half a billion board feet of Douglas fir in Oregon. Like the Big Blowup, Tillamook was merely the flagship of an armada of fires—Matilija, Selway, and others—that pummeled the West for three years. They prompted the Forest Service to adopt its famous 10:00 A.M. control policy, which meant that all fires were to be brought under control, if possible, by 10:00 the next morning. The policy would have been complete hubris were it not for a growing arsenal of fire-fighting tools: bulldozers that did the work of fifty men; airplanes that dropped thousands of gallons of retardant at a time; smoke jumpers who hit remote fires that would have taken men days to reach by foot. The idea was to get into the mountains fast and control the fires while they were still small; if a fire got away from the initial attack crews, thousands of lesser-trained men could be brought in to take over. As a tactic it made sense; as a public relations ploy it was unparalleled. Funding from Washington became effectively unlimited and remains so to this day.

As weather forecasting and communications improved, the loss of life declined, but mass tragedies still occurred. In 1937 a Civilian Conservation Corps crew of fourteen on the Blackwater fire were trapped between a spot fire and the main fire; they decided to turn and fight the spot rather than run, and they all died. In 1943 eleven marines died and seventy-two were injured when Santa Ana winds changed abruptly on the Hauser Creek fire in Southern California. In 1949 thirteen smoke jumpers and another man died in knee-high grass during a blowup on the Mann Gulch fire in Montana. In 1956 eleven convicts died on the Inaja fire in Southern California, and ten years later a hotshot crew lost twelve men in nearly identical circumstances on the Loop fire in the Angeles National Forest. A spot fire started at the bottom of a canyon, blew up unexpectedly, and ran twenty-two hundred feet of the canyon in less than a minute. That's twenty-five miles an hour. The crew didn't have a chance.

By the 1970s fire crews had portable fire shelters and Nomex clothing and could theoretically survive some burnovers. Fire shelters start to disintegrate at around six hundred degrees, though, and a fire run in heavy timber can hit temperatures three times that high. The only thing that will save a crew in the face of such an inferno is to get out of the way before it hits, and to that end researchers at the Intermountain Fire Science Laboratory in Missoula, Montana, have developed mathematical models that predict—given certain fuel conditions, terrain type, and meteorological conditions—what a fire will do. These models have been programmed into computers and can be used in conjunction with satellite data to project fire growth on any fire anywhere in the United States. The incident command team on a fire can punch its location into a computer, along with topographical and meteorological information, and receive very specific information about likely fire behavior the following day: which canyons will burn out; which ridgelines will hold.

Still, no amount of computing power can predict exactly what a fire will do.

"Nothing we're doing today is more important than a human life," one incident commander said at a 6:00 A.M. briefing.

That sentence—more than fire behavior models, lightweight fire shelters, or advanced meteorology—explains why people no longer die in the terrible numbers they used to on wildfires in the United States.

In the middle of the afternoon we moved down off the ridgetops toward H-6. Helicopters were to lift us and the three crews down to the fire camp for the night. I was in no hurry to leave the mountains, but the Union, Negrito, and Smokey Bear crews were being replaced by type two outfits that hadn't spent the last two nights spiked out. We clustered around the helispot and then piled onto the helicopters with our line packs and were shuttled back to camp.

That night I heard that a fire had just leaped the huge Salmon River Canyon up by McCall; a fire had started north of the river and a backburn had been set by dropping flaming Ping-Pong balls from a helicopter. (The Ping-Pong balls are filled with potassium permanganate and ignited by antifreeze. A needle injects the antifreeze at the moment of release and they ignite after thirty seconds, usually after they hit the ground. The helicopter pilot had circled to check his work, and one of the balls had accidentally released over the south side of the river. That was all it took. Within hours the new fire was fifty acres in size and crowning through heavy timber.

I drove up to McCall the next morning to see if I could get on the fire, but by then it was so far out of control that no one was allowed near it. Crews were cutting indirect line miles ahead of the fire and cold-trailing their way southward from the river, but that was it. The town of Riggins was filled with smoke; there was a mushroom cloud of smoke pumping out of the canyon; the airfield at McCall was crammed with reserve helicopters. But I wanted to see real flame. I slept on a sandbar along the North Folk of the Salmon River and drove back to Boise the next day. There was another fire north of town, I was told. It was racing through the dead-brown hills, and hot-shots were getting pulled off Foothills to deal with it. It was bad, and it was moving fast.

There's a little river that runs through Boise, and from the cafés that line the river walk, I could look up and watch the mountains burn. A big head of smoke was pumping out of the hills to the north, and retardant-streaked air tankers were making nonstop runs to and from the tanker base at the Boise airport, south of town. According to dispatch, bulldozers were trying to save a subdivision off Highway 21, wind-driven flames were racing through the terrifically dry sagebrush and cheatgrass around Lucky Peak Reservoir, and fifty crews left over from the Foothills fire were waiting to go in. I picked up a pass for the roadblock and headed north up Highway 21, toward the smoke.

Manpower was so short that the roadblock was guarded by a middle-aged couple in lawn chairs. I showed them my letter, and they waved me through. Soon I was alone on the dirt road that led into the burn zone. A line of flame hung like a necklace along the parched flanks of the hills. Smoke had turned the sunset blood-red. After three or four miles there was a hand-written cardboard sign that read: AREA CLEARED @ 19:30 HOURS 9/2—U.S.F.S. Just beyond that was the fire. It had reached the road and was swirling around a utility line that continued on up into the hills. I stopped the car and got out, completely alone with the fire and the mountains and the huge dead sky. Ten-foot tongues of flame licked the guardrails and shot into the sky. The vegetation died loudly, as if in pain, popping and exploding in the thickening dusk.

I took a few photographs and then went back to my car. The fire was about to jump the road. It would eventually move up into some timber and end up torching over thirteen thousand acres. A house would burn down. The beautiful Leonard ranch would be saved—barely—by ground and dozer crews backed up by massive air attack. It was called the Dunnigan Creek fire. It was one of roughly one hundred thousand wildfires during the summer of 1992, and if you ask a hotshot if he's ever heard of it, chances are he'll say no. Rain put it out after a couple of days.

Author's Note

This essay was written in 1992, and since then there have been many changes in the way wildfire is fought in the United States. I wanted to alter the original work as little as possible, so it should be noted that—among other changes—the Boise Interagency Fire Center is now called the National Interagency Fire Center; fire fighters are paid more than $8.50 an hour; and portable computers are in common use for predicting fire behavior.

Most significant, however, the fatality statistics have changed. When I was in Idaho, it had been twenty-six years since more than half a crew had died in a single incident—the Loop fire in southern California, where twelve hotshots were overrun in a matter of minutes.

Tragically, in 1994, fourteen hotshots and smoke jumpers were over-run in similar circumstances on the South Canyon fire outside Glenwood Springs, Colorado. The South Canyon fire is now the deadliest fire since 1937. My account of that incident appears as "Blowup," the second essay in this collection.

Many hotshots I spoke with attributed the increasing danger of their job to severe drought conditions in the northern Rockies, as well as to decades of rigorous fire suppression. Both have contributed to a huge buildup of dead fuel in our nation's forests—fuel that ordinarily would have been cleared out by the small fires that regularly flare up in an unmanaged ecosystem. A disastrous fire season was inevitable, and in 2000 it finally happened. Eighty-five thousand wildfires burned nearly seven million acres across the United States. Sixteen people died, and fire suppression cost over one billion dollars.

It was the worst season ever. With the western drought continuing unabated and huge amounts of deadwood still choking many forests, fire behavior experts don't expect conditions to get better anytime soon.

BLOWUP: WHAT WENT WRONG AT STORM KING MOUNTAIN

1994

The main thing Brad Haugh remembers about his escape was the thunderous sound of his own heart. It was beating two hundred times a minute, and by the time he and the two smoke jumpers running with him had crested a steep ridge in Colorado, everyone behind them was dead.

Their coworkers on the slope at their backs had been overrun by flames that Haugh guessed were three hundred feet high. The fire raced a quarter mile up the mountain in about two minutes, hitting speeds of eighteen miles an hour. Tools dropped in its path were completely incinerated. Temperatures reached two thousand degrees—hot enough to melt gold or fire clay.

"The fire blew up behind a little ridge below me," Haugh said later. "People were yelling into their radios, 'Run! Run! Run!' I was roughly one hundred and fifty feet from the top of the hill, and the fire got there in ten or twelve seconds. I made it over the top and just tumbled and rolled down the other side, and when I turned around, there was just this incredible wall of flame."

Haugh was one of forty-nine firefighters caught in a wildfire that stunned the nation with its swiftness and its fury. Fourteen elite fire fighters perished on a spine of Storm King Mountain, seven miles west of Glenwood Springs, Colorado. They died on a steep, rocky slope in a fire initially so small that the crews had not taken it seriously. They died while cars passed within sight on the interstate below and people in the valley aimed their camcorders at the fire from garage roofs.

There were many other fire fighters on Storm King when Brad Haugh crested the ridge, yet he feared that he and the two men with him were the only ones on the mountain left alive. That thought—not the flames—caused him to panic. He ran blindly and nearly knocked himself unconscious against a tree. Fires were spotting all around him as the front of flames chased him. The roar was deafening; "a tornado on fire" was how he later described it. The light, he remembered, was a weird blood-red that fascinated him even as he ran.

The two smoke jumpers with him were Eric Hipke and Kevin Erickson. Hipke had been so badly burned the flesh was hanging off his hands in strips. Haugh paused briefly to collect himself, then led the two men about a hundred yards down the mountain, stopping only long enough to wrap Hipke's hands in wet T-shirts. As they started down again, the fire was spreading behind them at a thousand acres an hour, oak, pinyon, and juniper spontaneously combusting in the heat.

"I didn't have any nightmares about it later," said Haugh. "But I did keep waking up in the night very disoriented. Once I had to ask my girlfriend who she was."

The South Canyon fire, as it was called, ignited on Saturday, July 2, as a lightning strike in the steep hills outside Glenwood Springs. At first people paid it little mind because dry lightning had already triggered thirty or forty fires across the drought-plagued state that day; another wisp of smoke was no big deal. But this blaze continued to grow, prompting the Bureau of Land Management (BLM) district office in

Grand Junction to dispatch a seven-member crew on the morning of July 5 to prepare a helicopter landing site, designated H-1, and start cutting a fire line along a ridge of Storm King. At this point the blaze was cooking slowly through the sparse pinyon and juniper covering the steep drainage below. Glenwood Springs was visible to the east, and a pricey development called Canyon Creek Estates was a mile to the west. Interstate 70 followed the Colorado River one thousand feet below, and occasionally the fire fighters could see rafters in brightly colored life jackets bumping through the rapids.

The BLM crew worked all day, until chain saw problems forced them to hike down to make repairs. Replacing them were eight smoke jumpers from Idaho and Montana (eight more would be added the next morning) who parachuted onto the ridgetop to continue cutting fire line. They worked until midnight and then claimed a few hours' sleep on the rocky ground.

Just before dawn, on the morning of July 6, Incident Commander Butch Blanco led the BLM crew back up the steep slope. Arriving at the top, Blanco discussed strategy with the smoke jumper in charge, Don Mackey. At about the same time, the BLM office in Grand Junction dispatched one additional crew to the fire, the twenty-member Prineville Hotshots, a crack interagency unit from Oregon whose helmet emblem is a coyote dancing over orange flame.

The smoke jumpers had cleared another landing spot, H-2, on the main ridge, and around twelve-thirty in the afternoon, a transport helicopter settled onto it. The first contingent of the Prineville crew ran through the rotor wash and crouched behind rocks as the chopper lifted off to pick up the rest of the unit from below. They'd been chosen alphabetically for the first flight in: Beck, Bickett, Blecha, Brinkley, Dunbar, Hagen, Holtby, Johnson, and Kelso. Rather than wait for their crew mates, these nine hotshots started downslope into the burning valley.

The layout of Storm King Mountain is roughly north-south, with a central spine running from the 8,793-foot summit to H-2. Another half mile south along this ridge was the larger site, H-1. The fire had started on a steep slope below these cleared safe areas and was spreading slowly.

The strategy was to cut a wide firebreak along the ridgetop and a smaller line down the slope to contain the blaze on the southwestern flank of the ridge. Flare-ups would be attacked with retardant drops from choppers. If there were problems, crews could easily reach H-1 in five or ten minutes and crawl under their fire shelters—light foil sheets that resemble space blankets and deflect heat of up to six hundred degrees.

"It was just an ugly little creeper," the BLM's Brad Haugh said of the early stages of the fire. Every summer, fire fighters like Haugh put out thousands of blazes like this one all over Colorado; at this point there was no reason to think South Canyon would be any different.

The second half of the Prineville crew dropped onto H-2 around 3:00 P.M. and began widening the primary fire line. Two hundred feet below, Haugh was clearing brush with his chain saw on a 33 percent slope. That meant the ground rose one foot for every yard climbed, roughly the steepness of a sand dune. The grade near the top was closer to 50 percent. He wore bulky Kevlar sawyer's chaps and a rucksack loaded with two gallons of water weighing fifteen pounds, a folding knife, freeze-dried rations, and some toilet articles. He also carried a folding fire shelter and a Stihl 056 chain saw that weighed ten or twelve pounds. Even loaded down as he was, Haugh could probably have reached the ridgetop in less than one minute if he had pushed it, and H-1 in five or ten minutes. Wildfires rarely spread faster than one or two miles an hour, and the vast majority of fire fighters are never compelled to outrun them—much less fight to survive them. By conventional fire evaluation standards, Haugh was considered safe.

About three-thirty Haugh took his second break of the day. It was so hot he had already consumed a gallon of the water he carried. The fire was burning slowly in the drainage floor, and the crews fighting it—nine from the Prineville unit and twelve smoke jumpers—

were several hundred feet below him in thick Gambel oak, some of the most flammable wood in the West.

Around 3:50 Haugh and his swamper—a sawyer's helper who flings the cut brush off the fire line—were finishing their break when their crew boss announced they were pulling out. Winds were picking up from a cold front that had moved in a half hour earlier, and the fire was snapping to life. They were ordered to climb to the ridgetop and wait it out.

It's rare for an entire mountainside to ignite suddenly, but it's not unheard of. If you stand near H-2 and look several miles to the west, you can see a mountain called Battlement Mesa. In 1976, three men died there in a wildfire later re-created in a training video called *Situation #8.* Every crew member on Storm King would certainly have seen it. In *Situation #8,* a crew is working upslope of a small fire in extremely dry conditions. Flames ignite Gambel oak and race up the hill, encouraged by winds. The steep terrain funnels the flames upward, and fire intensity careens off the chart, a classic blowup. Four men are overrun, three die. The survivor, who suffered horrible burns, says they were never alerted to the critical wind shift—an accusation the BLM denied at the time. "It's a hell zone, really," said one Forest Service expert on Colorado's oak- and pinyon-covered hills. "It's one dangerous son of a bitch."

At about 4:00 P.M. high winds hit the mountain and pushed a wall of flames north, up the west side of the drainage. Along the ridge, the BLM crew and the upper Prineville unit began moving to the safety of H-1. Below them, Don Mackey ordered his eight jumpers to retreat up to a burned-over area beneath H-1. He then started cross-slope to join three other smoke jumpers deployed with the Prineville nine. Apparently, no one had advised them that the situation was becoming desperate. In the few minutes it took Mackey to join the twelve fire fighters, the fire jumped east across the drainage. "I radioed that in," said Haugh. "And then another order came to evacuate." That order came from Butch Blanco on the ridgeline, who was hurriedly conducting the evacuation. "This was a much stronger warning than the previous one," recalled Haugh. "I sent my swamper to the ridgetop with the saw and radioed that as soon as the lower Prineville contingent came into sight below me, I would bump up to the safe zone."

Suddenly, fierce westerly winds drove the fire dangerously close—though still hidden behind the thick brush—to the unsuspecting fire fighters. "The crew was unaware of what was behind them," said Haugh. "They were walking at a slow pace, tools still in hand and packs in place." As Haugh watched them, a smoke jumper appeared at his side. "He said that his brother-in-law was down in the drainage, and he wanted to take his picture."

That fellow was Kevin Erickson, and Don Mackey was his brother-in-law, now in serious trouble below. As Erickson aimed his camera, everything below him seemed to explode. "Through the viewfinder, I saw them beginning to run, with fire everywhere behind them," Erickson said. "As I took the picture, Brad grabbed me and turned me around. I took one more look back and saw a wall of fire coming uphill." Closing in on Haugh and Erickson were smoke jumper James Thrash and the twelve other fire fighters in a ragged line behind him. Though Blanco and others were now screaming, "Run! Run! Run!" on the radio, Thrash chose to stop and deploy the fire shelter he would die in. Eric Hipke ran around him and followed Haugh and Erickson up the hill. The three-hundred-foot-high flames chasing them sounded like a river thundering over a waterfall.

In his book *Young Men and Fire,* Norman Maclean writes that dying in a forest fire is actually like experiencing three deaths: first the failure of your legs as you run, then the scorching of your lungs, finally the burning of your body. That, roughly, is what happens to wood when it burns. Water is driven out by the heat; then gases are superheated inside the wood and ignited; finally, the cellulose is consumed. In the end nothing is left but carbon.

This process is usually a slow one, and fires that burn more than a few acres per hour are rare. The South Canyon fire, for example, only burned fifty acres in the first three days. So why did it suddenly rip through two thousand acres in a couple of hours? Why did one hillside explode in a chain reaction that was fast enough to catch birds in midair?

Fire typically spreads by slowly heating the fuel in front of it—first drying it, then igniting it. Usually, a walking pace will easily keep fire fighters ahead of this process. But sometimes a combination of wind, fuel, and terrain conspires to produce a blowup in which the fire explodes out of control. One explanation for why South Canyon blew up—and the one most popular in Glenwood Springs—was that it was just so damn steep and dry up there and the wind blew so hard that the mountain was swept with flame. That's plausible; similar conditions in other fires have certainly produced extreme fire behavior. The other explanation turns on a rare phenomenon called superheating.

Normally, radiant heat drives volatile gases—called turpines— out of the pinyon and juniper just minutes before they are consumed. But sometimes hot air rises up a steep slope from a blaze and drives turpines out of a whole hillside full of timber. The gases lie heavily along the contours of the slopes, and when the right combination of wind and flame reaches them, they explode. It's like leaving your gas stove burners on for a few hours and then setting a match to your kitchen.

A mountainside on the verge of combustion is a subtle but not necessarily undetectable thing; there are stories of crews pulling out of a creepy-feeling canyon and then watching it blow up behind them. Turpines have an odor, and that's possibly why some of the Prineville survivors said that something had "seemed wrong." The westward-facing hillside had been drying all afternoon in the summer sun. Hot air was sucked up the drainage as if it were an open flue. The powerful winds that hit around 4:00 P.M. blew the fire up the drainage at the hottest time of day. And turpines, having baked for hours, could conceivably have lit the whole hillside practically at once.

When Storm King blew, Haugh had to run 150 feet straight up a fire line with poor footing. Despite rigorous conditioning—he is a runner and a bodybuilder—his heart rate shot through the roof and his adrenal glands dumped enough epinephrine into his system to kill a house cat. Behind him, sheets of flame were laid flat against the hillside by 50 mph winds. The inferno roared through inherently combustible vegetation that had been desiccated, first by drought, then by hot-air convection, finally by a small grass fire that flashed through a few days earlier. The moisture content of the fine dead fuels was later estimated to be as low as 2 or 3 percent—absolutely explosive. As Haugh ran, panicked shouts came over the tiny radio clipped to his vest for people to drop their equipment and flee. One brief thought flashed through his mind—"So this is what it's like to run for your life"—and he didn't think again until he'd reached the ridgetop.

Above him, the BLM and upper Prineville crews had abandoned hope of reaching H-1 and scrambled north toward H-2. When that route too was blocked, they turned and plunged over the ridge. Due south, one hundred feet below H-1, the eight smoke jumpers who had been ordered out by Don Mackey fifteen minutes earlier were crawling under their foil shelters to wait out the approaching fire storm. At Canyon Creek far below, a crew of fresh smoke jumpers who were preparing to hike in watched in horror as eight little silver squares appeared on the mountainside. Meanwhile, hidden from view by smoke, Mackey, the Prineville nine, and the three smoke jumpers were running a race only one of them, Hipke, would win.

In the end twelve of the dead were found along the lower fire line. Prineville hotshot Scott Blecha had also run past Thrash but lost his race a hundred feet from the ridgeline. The rest were in two main groups below a tree—*the* tree, as it came to be known, where Haugh had started his run—a few clumped so close together that their bodies were actually touching. Only smoke jumpers Thrash and Roger Roth had deployed their shelters, but the blistering heat disintegrated the foil. Kathi Beck died alongside Thrash, partly under his shelter. It seemed that in his last agony, Thrash may have tried to pull her in. In addition, Richard

Tyler and Robert Browning, two fire fighters deployed earlier to direct helicopter operations, perished just north of H-2, only a few hundred feet from a rocky area that might have saved them.

The Prineville nine's dash for safety ended after three hundred feet. They were caught just three or four seconds before Haugh himself cleared the ridgetop, and he could hear their screams over his radio. Reconstructing the details of the victims' agonized last seconds would occupy many hours of professional counseling for the survivors.

Dying in a fire is often less a process of burning than of asphyxiation. Their suffering was probably intense but short-lived. Pathologists looked for carbon in their lungs and upper airways and found none, which meant the victims weren't breathing when the fire passed over them. Their lungs were filled with fluid, their throats were closed in laryngeal spasms—responses to superheated air—and their blood contained toxic levels of carbon monoxide. This gas, given off during incomplete combustion, displaces oxygen in the blood and kills very quickly.

"They died after a few breaths at most," said Rob Kurtzman, a pathologist at the Grand Junction Community Hospital, "probably in less than thirty seconds. All the body changes—the charring, the muscle contractions, the bone fractures—happened after they were dead."

About four-thirty Haugh, Erickson, and Hipke staggered onto Interstate 70. Just an hour before, they had enjoyed a well-earned break on the mountain; now fourteen people were dead. But all they knew at that point was that Blanco, the incident commander, was calling out names on the radio and a lot of people weren't answering.

Haugh and Erickson laid Hipke in the shade of a police cruiser and doused him with water to lower his body temperature and prevent him from going into shock. Blanco climbed back up toward the fire to look for more survivors but found none. The eight smoke jumpers who'd deployed their shelters below H-1 emerged, shaken but unhurt.

They were saved not by their shelters but by having deployed them on previously burned ground. The fire was still pumping at this point, and Glenwood Springs was now in danger. Flames were racing eastward along the upper ridges, and the BLM command post at nearby Canyon Creek had begun ordering residents to evacuate.

Haugh's BLM crew had survived. The other Prineville Hotshots—the upper placements—made it out as well. They had snaked their way down the east side of the ridge through a hellish maze of spot fires and exploding trees. Two of them had tried to deploy their shelters but were dragged onward by friends.

Word quickly filtered back to BLM officials in Grand Junction that something terrible had happened on Storm King. Mike Mottice, the agency's area manager, had driven past the blowup and arrived at his Glenwood Springs office around 5:00 P.M. Minutes later crews began arriving from the mountain, and Mottice realized for the first time that there were people unaccounted for. "I hoped that the fire shelters would save them," he said. "But that evening some smoke jumpers confirmed that there were deaths."

The surviving Prineville crew members suspected that some had died, but they didn't know for sure until later that night. They were shuttled first to the Glenwood Springs office, then to Two Rivers Park at the center of town. An open-air concert was in progress, and they sat in their fire clothes while the mountains burned and local youths took in the music. Finally, around nine, a social worker named Carol Kramer arrived with Prineville crew boss Brian Scholz. Kramer was to take the crew back to the Ramada Inn. A conference room was quickly prepared where she could tell them privately that nine of their friends had died.

"When we reached the hotel, they started falling apart," said Kramer. "At that point, they knew. They were begging us to tell them, to just get it over with. I told them it was bad, that twelve were dead and five were missing."

The survivors' reaction was quick and violent. Some sobbed; others pounded tables. One fire fighter fled the room and threw up. Two crew members quickly left, followed by Scholz, who wanted to keep an eye on them. As a crew boss Scholz considered himself still on duty, and he refused to lose control in front of his men. Gradually a list of survivors was compiled.

"For a while there was a lot of being out of control," said Kramer. "Then for a few hours the sobbing was only intermittent; finally, there were a lot of thousand-yard stares. They'd just sit together silently. The next morning they ate a little food. It was a small thing, but that's what you look for."

Some of the most traumatized accepted individual counseling. One thing they needed was to describe the things they had experienced. One man relayed in excruciating detail the sounds of screams and shouts he had heard as he escaped over the hill. Within thirty-six hours, the eleven Prineville survivors were flown home to Oregon— in part to reunite them with their families, in part to protect them from being hounded by the national press corps. The Ramada Inn had become a shark pool of competing journalists, and the last thing survivors needed was TV cameras panning their faces for tears and anguish.

On Monday, July 11, a memorial service was held in Glenwood Springs. While Storm King Mountain smoldered in the background, helicopters flew in formation overhead and people wept to the strains of "Amazing Grace." President Clinton called Governor Roy Romer from Air Force One, and flags on government buildings throughout the country were at half-mast. At dawn the next day, the bodies of the Prineville nine were driven to Walker Field in Grand Junction and then flown home in a Forest Service DC-3. The remains were delivered to four different airfields in Oregon while honor guards played taps and next of kin received the caskets on the tarmac.

Before the embers were even cold on Storm King Mountain, a ten-member investigation team was convened and given forty-five days to examine the site and deliver its findings. The team was composed of former fire fighters and experts in fire behavior, meteorology, and safety equipment. The question of specific blame, however, was not supposed to arise; it was to be a strictly analytical study of what had happened and when.

The preferred view among most federal fire personnel and even most South Canyon survivors was that the West was apocalyptically dry and huge fires were bound to happen. On such fires, people sometimes die; indeed, there are a few fatalities every year. "I would go out on a fire line again with any person who was there," insisted the BLM's Mike Hayes. "We were doing the best we could with the resources we had. I mean, there were fifty fires in our district at the time."

A siege mentality developed in Glenwood Springs. Questions of specific culpability were construed as lack of respect for the fire fighters and even for the dead. On Monday, July 11, the *Glenwood Post* ran an article titled "Glenwood Incident Commander: Plans for Escape Worked," a daring stance to adopt concerning a fire in which fourteen people died. Butch Blanco had told the *Grand Junction Daily Sentinel* the day before that one smoke jumper (Hipke) who'd escaped had started his run behind the ill-fated Prineville crew. That suggested to him that there had been sufficient time to reach a safe area ahead of the fire. "Whether they [the Prineville nine] didn't take it seriously, I don't know," he said.

The first people to see the dead were the smoke jumpers who had deployed fire shelters below H-1. "I walked straight to the lower group of bodies and called for a helicopter," said smoke jumper Anthony Petrilli. "They asked if we needed medevac. I told him it was too late for that, and then I walked up the hill and found six more."

An hour later twenty-six smoke jumpers helicoptered in to investigate further. It was an early, unnatural dusk on the mountain as they picked their way past the charred bodies. They reported eleven dead and three missing. Within an hour, Governor Romer was on the scene; he told the smoke jumpers he wanted to remove the bodies as quickly as possible. The jumpers objected, saying that this was no different from a crime scene and the bodies should be left until someone examined them. Romer abided by their wishes. The next morning investigators began to measure things, ponder the dynamics of the mountain, and coax secrets from the dead.

The first question was how fast the fire had moved, and Haugh's estimate—that the last three hundred feet were covered in about twelve seconds—turned out to be close. In the end, the investigators confirmed that the fire had covered the quarter-mile slope in about two minutes, hitting its top speed of 18 mph in the dried-out Gambel oak.

The next question was why it had done that. Fire behavior is determined by an incredibly complicated interaction of fuel, terrain, and wind, and there are mathematical models describing the interaction. (The models are programmed into hand-held calculators carried by most incident commanders these days.) The deadly hillside faced west at a 33 to 50 percent slope, and the vegetation on it possessed burning characteristics described in a formula called Fuel Model Number Four. The moisture content of the small dead fuels on Storm King Mountain was around 3 percent. And the live Gambel oak (which had only been partly burned earlier) was several times drier than normal. In a light wind, according to this model, those conditions would produce twenty-three-foot flames spreading at a maximum of seven hundred feet an hour.

That's a manageable fire, or at least one that can be outrun, but an increase in wind speed can change the situation dramatically. At 7:20 P.M. on Tuesday (less than twenty-four hours before the blowup), the National Weather Service issued a "Red Flag" fire warning for the area around Glenwood Springs. Dry thunderstorms were expected the following morning, followed by southwest winds gusting up to 30 mph. A cold front would come through sometime that afternoon, swinging the winds to the northwest.

Gusts of 35 mph, plugged into Fuel Model Number Four, produce sixty-four-foot flames racing up the mountain at up to fifteen feet per second. In the superdry Gambel oak, the rate of spread would have been almost twice that—much faster than any human can run. The lessons of the Battlement Mesa fire (detailed in the *Situation #8* video) had not been learned: A small fire on steep ground covered with extremely dry vegetation had once more exploded in a mathematically predictable way—again, with tragic results.

The 226-page federal investigators' report concluded that just about everyone involved had been negligent in some way. Ground crews had been arrogant about the fire danger; supervisors had ignored local fuel and drought conditions; and the Western Slope Fire Coordination Center had failed to relay crucial weather information to the fire crews in the field. "Extreme fire behavior could have been predicted by using weather forecasts and information readily obtainable at the BLM Grand Junction District Office," read one of many such findings.

The most horrifying conclusion of the report was that twelve of the victims could have easily escaped from the valley if they had started running when evidence of extreme danger first emerged. Instead, they began a slow walk, some of them dying with their tools in their hands. This meant two things: The order for an all-out retreat was given far too late, and the victims had an inherent reluctance to acknowledge the seriousness of their situation. "Putting down the saw jacked the pucker factor up one notch," said smoke jumper Petrilli, who himself had not accepted the fact that he was running for his life until he put down his tools. The last thing fire fighters are supposed to do is give up a saw or shovel, so they are understandably loath to do so, since it means they are in a life-threatening situation.

"I know in my heart," said Haugh, "that the twelve persons who died in that part of the fire were unaware of what was happening." By the time the Prineville nine and the three smoke jumpers with them saw the horror coming—by the time great sheets of flame hit the dry Gambel oak and frantic voices over the radio screamed at them to run—they had only twenty seconds to live. They must have died in a state of bewilderment almost as great as their fear.

THE WHALE HUNTERS

1995

The last living harpooner wakes to the sound of wind. It has been blowing for two weeks now, whipping up a big ugly sea, ruining any chance of putting out in the boat. On this strong, steady wind, the northeast trades, European slave ships rode to the New World bringing fifteen million Africans across the Atlantic. One of their descendants now creeps through his house in the predawn gloom, wishing the wind would stop.

The man's name is Athneal Ollivierre. He is six feet tall, seventy-four years old, straight and strong as a dock piling. His hair rises in an ash gray column, and a thin wedge of mustache suggests a French officer in the First World War. On his left leg, there's the scar of a rope burn that went right down to the bone. His eyes, bloodshot from age and the glare of the sun, focus on a point just above my shoulder and about five hundred miles distant. In the corner of his living room rests a twenty-pound throwing iron with a cinnamon-wood shaft.

Ollivierre makes his way outside to watch the coming of the day. The shutters are banging. It's the dry season; one rainfall and the hills will be so covered with poui flowers that it will look as if it had just snowed. Shirts hang out to dry on the bushes in front of his house, and a pair of humpback jawbones forms a gateway beyond which sprawls the rest of his world, seven square miles of volcanic island that drop steeply into a turquoise sea. This is Bequia, one of thirty-two islands that make up the southern Caribbean nation of St. Vincent and the Grenadines. Friendship Bay curves off to the east, and a new airport, bulldozed across the reefs, juts off to the west. More and more tourists and cruise ships have been coming to Bequia, the planes buzzing low, the gleaming boats anchoring almost nightly in the bay, but at the moment that matters very little to Ollivierre. He's barefoot in the tropical grass, squinting across the water at a small disturbance in the channel. Through binoculars it turns out to be a wooden skiff running hard across the channel for the island of Mustique. It emerges, disappears, emerges again behind a huge green swell.

"Bequia men, they brave," he says, shaking his head. He speaks in a patois that sounds like French spoken with an Irish brogue. "They brave too much."

Ollivierre hunts humpback whales from a twenty-seven-foot wooden sailboat called the *Why Ask*. As far as he's concerned, his harpooning days are over, but he's keeping at it long enough to train a younger man, forty-three-year-old Arnold Hazell, to do it. Otherwise the tradition—and the last remnant of the old Yankee whaling industry—will die with him. When they go out in pursuit of a whale, Ollivierre and his five-man crew row through the surf of Friendship Bay and then erect a sail that lets them slip up on whales undetected. Ollivierre stands in the bow of his boat and hurls a harpoon into the flank of an animal that's five hundred times as heavy as he is. He has been knocked unconscious, dragged under, maimed, stunned, and nearly drowned. When he succeeds in taking a whale, schools on Bequia are let out, businesses are closed, and a good portion of the forty-eight hundred islanders descends on the whaling station to watch and help butcher, clean, and salt the whale.

"It's the only thing that bring joy to Bequia people," says Ollivierre, a widower whose only son has no interest in whaling. "Nobody don't be in their homes when I harpoon a whale. I retired a few years ago, but the island was lacking of the whale, and so I go back. Now I'm training Hazell. When I finish with whalin, I finish with the sea."

When a whale is caught, it's towed by motorboat to a deserted cay called Petit Nevis and winched onto the beach; the winch is a rusty old hand-powered thing bolted to the bedrock. Butchering a forty-ton animal is hard, bloody work—work that has been condemned by environmentalists around the world—and the whalers offer armloads of fresh meat to anyone who will help them. Some of the meat is cooked right there on the beach (it tastes like rare roast beef), and the rest is kept for later. The huge jawbones are sold to tourists for around a thousand dollars, and the meat and blubber are divided up equally among the crew. Each man sells or gives his share away as he sees fit. "Who sell, sell; who give, give," as Ollivierre says. The meat goes for two dollars a pound in Port Elizabeth.

If there is a species that exemplifies the word *whale* in the popular mind, it's probably the humpbacks that Ollivierre hunts. These are the whales that breach for whale-watching boats and sing for marine biologists. Though nearly 90 percent of the humpback population has been destroyed in the last hundred years, at least half of the remaining eleven thousand humpbacks spend the summer at their feeding grounds in the North Atlantic and then migrate south in December. They pass the winter mating, calving, and raising their young in the warm Caribbean waters, and when the newborns are strong enough— they grow a hundred pounds a day—the whales journey back north.

It is by permission of the International Whaling Commission (IWC) based in Cambridge, England, that Ollivierre may take two humpbacks a year. In 1986 a worldwide moratorium was imposed on all commercial whaling, but it allowed "aboriginal people to harvest whales in perpetuity, at levels appropriate to their cultural and nutritional requirements." A handful of others whale—in Greenland, Alaska, and Siberia—but Ollivierre is the only one who still uses a sailboat and a hand-thrown harpoon. These techniques were learned aboard Yankee whaling ships a hundred years ago and brought back to Bequia without changing so much as an oarlock or clevis pin.

"You came and put a piece of your history here, and it's still here today," says Herman Belmar, a local historian who lives around the corner from Ollivierre. Belmar is a quiet, articulate man whose passion is whaling history. He is trying to establish a whaling museum on the island. "Take the guys from Melville's *Moby-Dick* and put them in Athneal's boat, and they'd know exactly what to do."

One day at dawn I drive over to meet Ollivierre. His house is a small, whitewashed, wood and concrete affair on the side of a hill, surrounded by a hedge. Except for the whalebone arch, it's indistinguishable from any other house on the island. I let myself through a little wooden gate and walk across his front yard, past an outboard motor and a vertebra the size of a barstool. It's mid-February, whaling season, and Ollivierre is seated on a bench looking out across the channel. I stick out my hand; he takes it without meeting my eye.

By Bequia standards, Ollivierre is a famous man. Many people have stood before him asking for his story, but still I'm a little surprised by his reaction. Not a word, not a smile—just the trancelike gaze of someone trying to make out a tiny speck on the horizon. I stand there uncomfortably for a few minutes and finally ask what turns out to be the right question: "Could I see your collection?"

If you wander around Port Elizabeth for any length of time, a taxi driver will inevitably make you the offer: "Come meet the real harpooner! Shake his hand, see his museum!" A museum it's not, but Ollivierre has filled the largest room of his house with bomb guns, scrimshaw, and paintings. The paintings are by a local artist and commemorate some of Ollivierre's wilder exploits: *Athneal Done Strike de Whale,* reads one. As Ollivierre discusses his life, he slowly becomes more animated and finally suggests that I walk up to the hilltop behind his house to meet the rest of the crew.

A path cuts up the hill past another low wood and concrete house.

Split PVC pipe drains the roof and empties into a big concrete cistern, which is almost dry. (Every drop of drinking water on Bequia must be caught during the rainy season.) At the top of the hill are some wind-bent bushes and a thatch and bamboo sunbreak that tilts south toward the sea. Four men sit beneath it, looking south across the channel. They gnaw on potatoes, pass around binoculars, suck on grass stems, watch the sky get lighter. In the distance is a chain of cays that used to be the rim of a huge volcano, and seven miles away is the island of Mustique. When the wind permits, the whalers sail over there to look out for humpbacks.

"Hello. Athneal sent me," I offer a little awkwardly.

The men glance around. There's been some bad press about whaling, even the threat of a tourist boycott, and everyone knows this is a delicate topic. An old man with binoculars motions me over. "We can tell whatever you want," he says, "but we can't do anything without Dan, de cop'm."

After Ollivierre, Dan Hazell, who bears some distant relation to Arnold Hazell, is the senior member of the crew. He's the captain, responsible for maneuvering the boat according to Ollivierre's orders. A young man named Eustace Kydd says he'll round up Dan and a couple of others and meet me at a bar in Paget Farm. Paget Farm is a settlement by the airport where the whalers live: ramshackle houses, dories pulled up on rocks, men drinking rum in the shade. Most of the men on the island make their living net fishing. I nod and walk back down the hill. Ollivierre is still in his yard, glassing the channel and talking to a young neighborhood man who has dropped by. They give me a glance and keep talking. The wind has dropped; the sun is thundering impossibly fast out of the equatorial sea.

Unfortunately for Ollivierre, the antiquity of his methods has not exempted him from controversy. First of all, he has been known to take mother-calf pairs, a practice banned by the IWC. In addition, Japan started giving St. Vincent and several neighboring islands tens of millions of dollars in economic aid after the imposition of an international moratorium on whaling in 1986. The aid was ostensibly to develop local fisheries, but American environmental groups charged that Japan was simply buying votes on the IWC. The suspicions were well founded: St. Vincent, Dominica, and Grenada have received substantial amounts of money from Japan, and all have voted in accordance with Japan's whaling interests over and over again.

Things came to a head last year when the IWC introduced a proposal to create an enormous whale sanctuary around Antarctica. The sanctuary would offer shelter to whales as the worldwide moratorium was phased out in keeping with growing whale populations. The Massachusetts-based International Wildlife Coalition, headed by Dan Morast, threatened to organize a tourist boycott against any country that voiced opposition to the proposal, and in the end only Japan voted against it. St. Vincent, Dominica, and Grenada abstained from the vote, and the South Seas Sanctuary was passed.

But the controversy over Bequia is more emotional than a vote. Ollivierre has become the focal point for dozens of environmental lobbyists, for whom everything he does is drenched in symbolism. First there was Ollivierre's flip-flop: In 1990 he announced his retirement, but a year later he was back at it, sitting on his hilltop, looking out for whales. It was a move that angered environmentalists who thought they'd seen the last of whaling on Bequia. The reaction was compounded by Ollivierre's efforts to sell the island of Petit Nevis, the tiny whaling station that has belonged to his family for three generations; a Japanese businessman's offer of five million dollars was an outrage. Of course, Ollivierre's personal impact on the humpback population is negligible. Morast's point seems to be more conceptual: that the land sale is just another form of bribery to encourage the St. Vincent representative on the IWC to vote for whaling.

And contrary to Morast's view, Ollivierre would love to retire. His joints ache; his vision is clouded; he's an old man. Harpooning is dangerous, and apprentices are hard to come by. Several years ago he trained his nephew Anson Ollivierre to harpoon, but Anson branched out on his own before even bloodying his hands. Now he's building his own whaleboat, and Ollivierre fears Anson will get his whole crew killed. So this year Ollivierre tried again, taking on Arnold Hazell. Hazell's great-grandfather crewed for Ollivierre's great-grandfather, and now, a hundred years later, the relationship continues. Since there are no whales to practice on, Hazell just hangs out at Ollivierre's house, listens to the old stories, soaks up the lore.

When Hazell has killed his first whale, Ollivierre will retire. And the antiwhaling community will have a new face upon which to hang its villain's mask.

A short time after meeting with Ollivierre and his crew, I drive down to Paget Farm to see about going out on the whaleboat. On the way I pass a new fish market, paid for by the Japanese government as part of a six-million-dollar aid package. According to the Japanese, it's a no-strings-attached token of their affection for the Bequia fishermen. Past the market I turn onto a narrow cement road that grinds up a desperately steep hillside. At one end of the road is the sky; at the other end is the sea. The appointed bar is a one-story cement building halfway up the hill. I park, chock the wheels, and wander inside. It's as clean and simple inside as out: a rough wooden counter, a half dozen chairs, no tables, a big fan. The walls are a turquoise color that fills the room with cool coral reef light. A Save the Whales poster hangs in tatters on one wall, and a monumental woman opens soft drinks behind the counter. Five men are ranged at the far end of the room. They are dressed in T-shirts and baggy pants, and one has a knife in his hand. Captain Dan, too shy to speak, just looks out the window into the midday heat. Arnold Hazell greets me with a smile and makes his pitch.

"In Bequia we don't have much opportunity like you in the

States," he begins. "We grew up on the sea an live from the sea. Even if we don't cotch a whale for the next ten years, it will be good just to be whalin. Just to keep the heritage up. Japanese an Norwegians—they killin whales by the thousands, an those people could afford to do something else. They have oil; they have big industry; they have a better reason to stop." He pauses. "You know, we can put the boat out, we can talk to you, you can take snaps, but it a whole day's work for us. We need something back."

Luckily, I've been told about this ahead of time. It's a tourist economy—the sunshine, the water, the beaches, it's all for sale—and the whalers see no reason why they should be any different. A young man in dreadlocks steps in quietly and leans against the bar. He listens with vague amusement; he's heard this all before.

"A few years ago a French crew come here," says Eustace. "They come to make a film. They offer us thousands of dollars: they pre-pared to pay that. But we say no because we know they makin so much more on the film. Why should we work an they make all the money?"

After this statement, negotiations proceed slowly. Some careful wording, a few ambiguous phrases, and finally an agreement is reached: We'll meet at Friendship Bay tomorrow before dawn. "And," says Captain Dan, his eyes never wavering from the horizon, "you'll see the *Why Ask* fly."

In the distant past, most of the Caribbean islands were inhabited by the peace-loving Arawak people. Very little is known about them, be-cause most were killed, and the rest were driven from the islands by the Caribs, whose name comes from the Arawak word for "cannibal." Unfortunately for the Caribs, Columbus discovered their bloody lit-tle paradise within years of their ascendancy, and two hundred years later most of them were gone as well. Bequia—dry, tiny, and poor— was one of their last hideouts, and when the French finally settled here, they found people of mixed Carib-African ancestry hiding in the hills. The Africans, as it turned out, had swum ashore from a wrecked slave ship, the *Palmira,* in 1675.

France ceded Bequia to Britain in 1763, and inevitably the Black Caribs, as they were called, were put to work on the local sugar and cotton plantations. Only free labor could coax a profit from such poor soil, and when the British abolished slavery in 1838, Bequia's economy fell apart. The local elite fled, and islanders reverted to farming and fishing—and eventually whaling—to survive.

The first Bequian to kill a whale was Bill Wallace, a white landowner's son who went to sea at age fifteen and returned twenty years later with a New England bride and an armful of harpoons. As a child on Bequia he'd watched humpbacks spouting offshore during the winter months, and he didn't see why boats couldn't put out from the beach to kill them. Crews could keep lookout from the hilltops and then man their boats when they saw a spout. He recruited the strongest young men he could find and established the first whaling station on Friendship Bay in 1875.

There was nothing benevolent in Wallace; he was a tough old salt who was essentially out for his own gain. He'd lost his father shortly before leaving the island and had grown up in an industry that was considered brutal even by the standard of the times. Whaling crews were at sea for three or four years at a stretch, under conditions that would have made prisoners of war balk. Captains had absolute authority over their men, and there were instances of errant crew members being whipped to death. The crews themselves were no blessing, often largely composed of criminals, drunks, and fresh-faced kids just off the farm. It's easy to guess whose habits, after four years at sea, rubbed off on whom.

The only thing that kept such an enterprise together was the unspeakable danger these men faced and the financial rewards of making it through alive. The largest whales in the world—blue whales—weigh 190 tons and measure up to one hundred feet long. They have

hearts as big as oil drums and penises nine feet long. In 1839 an enraged sperm whale, immortalized by Herman Melville in *Moby-Dick,*
attacked a three-masted ship and sank her. Harpooning such a creature could be compared with standing on a highway overpass and
harpooning a truck. The line runs out through the wooden chocks so
fast that it can catch fire. At that speed a kink in the line could catch
a man's arm and rip it out of the socket or make the man simply disappear from the boat. A slack line is even worse: It means the whale
has sounded and is coming straight back up toward the bottom of the
boat. The terror inspired by whales affected men differently. Some,
mesmerized by the risks and the money, whaled until they grew old or
were killed. Others jumped out of the longboats at their first sight of
their foe.

Four in the morning, the air soft as silk. I'm speeding along the dark
roads in a rented jeep, slowing down just enough to survive the speed
bumps. The northern part of Bequia is almost completely uninhab-
ited, steep, scrub-choked valleys running up to cliffs of black volcanic
rock. Shark Bay, Park Bay, Brute Point, Bullet. Between the head-
lands are white-sand beaches backed by cow pastures and coconut
groves. The road passes a smoldering garbage dump, climbs the island's
central ridge, and then curves into Port Elizabeth. The only signs of
life at this hour are a few dockworkers loading a rusted interisland
cargo ship under floodlights. The road claws up a hill and then crests
the ridge above Lowerbay—Lowby, as it's called—and starts down
toward Friendship.

A dry wind is blowing through the darkness, and the surf against
Semples Cay and St. Hilaire Point can be heard a mile away. I pull off
the road near Ollivierre's house and feel my way down a steep set of
cement stairs to the water's edge. The surf smashes white against the
outer reefs; everything else is the blue-black of the tropics just before
dawn. The whalers arrive ten minutes later, as promised, moving sin-

gle file down the beach. They stow their gear without a word and put their shoulders to the gunwales of the *Why Ask;* she rolls heavily over four cinnamonwood logs and slips into the sea. The wind has abated enough to sail to the preferred lookout on Mustique; otherwise we'd have to make do with the hill above Ollivierre's.

Within minutes they're under way: Captain Dan at the tiller, Ollivierre up front, and Biddy Adams, Eustace, Arnold Hazell, and Kingsley Stowe amidships. They pull at the eighteen-foot oars, plunging into the surf. Once clear of the reef they step the mast, cinch the shrouds, becket the sprit and boom. They scramble to work within the awkward confines of the boat as Ollivierre barks orders from the bow.

The *Why Ask* is heartbreakingly graceful under sail, as much a creature of the sea as the animals she's designed to kill. She was built on the beach with the horizon as a level and Ollivierre's memory as a plan. Boatwrights have used such phrases as "lightly borne" and "sweet-sheared and buoyant" to describe whaleboats of the last century, and they apply equally to the *Why Ask.*

The boat quickly makes the crossing to Mustique, where the crew spends half the day on a hilltop overlooking the channel. With an older whaler named Harold Corea stationed above Ollivierre's house with a walkie-talkie, they have doubled the sweep of ocean they can observe. In addition, they often get tips from fishermen, pilots, or people who just happen to look out their windows at the right moment. These people are always rewarded with whale meat if the chase is successful.

In the early days, between 1880 and 1920, there were nine shore whaling stations throughout the Grenadines, including six on Bequia, and together they surveyed hundreds of square miles of ocean. They'd catch perhaps fifteen whales in a good year, a tremendous boon to the local economy. In 1920, 20 percent of the adult male population of Bequia was employed in the whaling industry.

Five years later all that changed; a Norwegian factory ship set up operation off Grenada and annihilated the humpback population within a year and a half. Almost no whales were caught by islanders between 1925 and 1948, and none at all for eight years after that. The whaling stations folded one by one, and by the 1950s only the Ollivierre family was left. Today the humpback population has recovered slightly—the IWC now considers the species "vulnerable" rather than "endangered"—but sightings off Bequia are still rare. Last year the crew put out after a whale only once; so far this season they have yet to see a spout.

The boat returns from Mustique in the afternoon with nothing to report. The crew shrugs it off: Waiting is as much a part of whaling as throwing the harpoon.

On those lucky occasions when Ollivierre spots a whale from Mustique, he fixes its position in his mind, sails to the spot, and waits. If there's no wind, the crew is at the oars, pulling hard against oarlocks that have been lined with fabric to keep them quiet. Humpbacks generally dive for ten or fifteen minutes and then come up for air; each time they do, Ollivierre works the boat in closer. The harpoon, protected by a wooden sheath, rests in a scooped-out section of the foredeck called the clumsy cleat; when the harpoon is removed, it fits the curve of Ollivierre's thigh perfectly.

The harpoon is heavy and brutally simple. A thick cinnamon-wood shaft has been dressed with an ax and pounded into the socket of a throwing iron. The head itself is made of brass and has been ground down to the edge of a skinning knife; it is mounted on a pivot and secured by a thin wooden shear pin driven through a hole. Upon impaling the whale, the pin breaks, allowing the head to toggle open at ninety degrees, catching deep in the flesh of the whale. It's a design that hasn't changed in 150 years. The harpoon is attached to a nine-fathom nylon tether, which in turn is tied—"bent," Ollivierre says—to the manila main line, which is 150 fathoms long. The line passes through a notch in the bow, runs the length of the boat, takes two wraps around the loggerhead, and is coiled carefully into a wooden tub. The loggerhead is a hefty wooden block that provides enough friction to keep the whale from running out the entire tub of rope. When a whale is pulling the line, Eustace scoops seawater over the side and fills the tub; otherwise the friction will set the loggerhead on fire. Meanwhile, Ollivierre takes his position in the bow, delivering orders to Captain Dan in a low, harsh voice. Above all, they must stay clear of the tail: It's powerful enough to launch a humpback clear out of the water and could obliterate the boat in a second. Ollivierre's leg is braced against the clumsy cleat, and the other men are wide-eyed at the gunwales, the rank smell of whale vapor in their faces. The harpoon has been rid of its sheath, and Ollivierre holds it aloft as if his body has been drawn like a bow, right hand cupping the butt end, left hand supporting the shaft. You don't throw a harpoon; you drive it, unloading it downward with all your weight and strength the moment before your boat beaches itself—"wood to blackskin"—atop the whale.

"De whale make no sound at all when you hit it. It just lash de tail and it gone," says Ollivierre. "Dan let go of everything an put his two hands on de rope. De whale have to take de rope from him; he have to hold it down."

A struck whale gives a few good thrashes with its tail and then tries to flee. It is a moment of consummate chaos: the line screaming out through the bow chock; the crew trying to lower the mast; the helmsman bending the line around the smoking loggerhead. Some men freeze, and others achieve ultimate clarity. "After we harpoon it, that frightness, that cowardness go from me," says Harold Corea, who at sixty-three is one of the oldest members of the crew. "It all go away; I become brave, I get brave."

Brave or not, things can go very wrong. Around 1970—Ollivierre doesn't remember exactly when—a whale smacked the boat with a fluke, staving in the side and knocking Ollivierre out cold. When he came to, he realized that the rope had grabbed him and turned his leg into a loggerhead. It sawed down to bone in an instant, cauterizing the arteries as it went, and nearly ripped his hand in half. Ollivierre refused to cut the rope because he didn't want the whale to get away, but finally the barnacle-encrusted fluke severed it for him. The boat returned to shore, and Ollivierre walked up the beach unassisted, his tibia showing and his foot as heavy as cement. Two men on the beach fainted at the sight.

There is no such thing as an uneventful whale hunt; by definition it's either a disaster or almost one. As soon as the harpoon is fast in the whale, the crew drops the mast and Dan tightens up on the loggerhead to force the whale to tow the boat through the water, foredeck awash, men crammed into the stern, a twenty-knot wake spreading out behind. Too much speed and the boat will go under; too much slack and the whale will run out the line. (There is one account of a blue whale that towed a ninety-foot twin-screw chaser boat, its engines going full bore astern, for fifty miles before tiring.) Every time the whale lets up, the crewmen put their hands on the line and start hauling it back in. The idea is to get close enough for Ollivierre to use either a hand lance or a forty-five-pound bomb gun, whose design dates back to the 1870s. It fires a shotgun shell screwed to a six-inch brass tube filled with powder that's ignited by a ten-second fuse. Ollivierre packs his own explosives and uses them with tremendous discretion.

The alternative to the gun is a light lance with a rounded head that doesn't catch inside the whale; standing in the bow, Ollivierre thrusts again and again until he finds the heart. "De whole thing is dangerous, but de going in and de killing of it is de most dangerous," he says. He's been known to leap onto the back of the whale and sit with his legs wrapped around the harpoon, stabbing. Sometimes the whale sounds, and Ollivierre goes down with it; if it goes too deep, he lets go and the crew pulls him back to the boat. When his lance has found the heart, dark arterial blood spouts out the blowhole. The huge animal stops thrashing, and its long white flippers splay outward. Two men go over the side with a rope and tie up the mouth; otherwise water will fill the innards and the whale will sink.

As dangerous as it is, only one Bequian has ever lost his life in a whaleboat: a harpooner named Dixon Durham, who was beheaded by a whale's flukes in 1885. So cleanly was he slapped from the boat that no one else on board was even touched. The closest Ollivierre has come to being Bequia's second statistic was in 1992, when the line caught on a midship thwart and pulled his boat under. He and his crew were miles from Bequia, and no one was following them; Ollivierre knew that without the boat, they would all drown. He grabbed the bow and was carried down into the quiet green depths. Equipment was rising up all around him: oars, ropes, wooden tubs. He hung on to the bow and clawed desperately for the knife at his belt. By some miracle the rope broke, and the whole mess—boat, harpoons, and harpooner—floated back up into the world.

Ollivierre found his VHF radio floating among the wreckage and called for help. Several days later, some fishermen in Guyana heard a terrible slapping on the mudflats outside their village and went to investigate. They found Ollivierre's whale stranded on the beach, beating the world with her flippers as she died.

The next day, Ollivierre, Hazell, and Corea are back up at the lookout, keeping an eye on the sea. Corea, who was partially crippled by an ocean wave at age nineteen, is one of the last of the old whalers. Hazell is the future of Bequia whaling, if there is such a thing. They sit on the hilltop all morning without seeing a sign. No one knows where the whales are. A late migration? A different route? Are there just no more whales?

After a couple of hours Ollivierre is ready to call it quits for the day. If the others see a spout, they can just run over to his house and tell him. More than anything he just seems weary; he's whaled for thirty-seven years and fished up until a few years ago. Enough is enough. He says good-bye and walks slowly down the hill. Corea watches him go and scours the channel one more time.

Hazell squats on a rock in the shade with half his life still ahead of him. He is neither old nor young, a man caught between worlds, between generations. Down the hill is a scarred old man who's trying to teach him everything he knows; across the ocean is a council of nations playing tug-of-war with a twenty-seven-foot sailboat. Hazell would try to reconcile the two, if it were possible, but it's not. And so he's left with one simple task: to visualize what it will be like to face his first whale.

A long winter swell will be running. The sunlight will catch the spray like diamonds. He'll be in the bow with his thigh against the foredeck and the harpoon held high. The past and the future will fall away, until there are no politics, no boycotts, no journalists. There will be just one man with an ancient weapon and his heart in his throat.

ESCAPE FROM KASHMIR

1996

The guerrillas appeared on the ridgeline shortly before dusk and walked down the bare hillside into the Americans' camp without bothering to unsling their guns. They were lean and dark and had everything they needed on their persons: horse blankets over their shoulders; ammunition belts across their chests; old tennis shoes on their feet. Most of them were very young, but one was at least thirty and hard-looking around the eyes—"a killer," one witness said. Jane Schelly, a schoolteacher from Spokane, Washington, watched them come.

"There were ten or twelve of them," she says, "and they were dressed to move. They didn't point their guns or anything; they just told us to sit down. Our guides told us they were looking for Israelis."

Schelly and her husband, Donald Hutchings, were experienced trekkers in their early forties who took a month every summer to travel somewhere in the world: the Tatra Mountains in Slovakia; the Annapurna Massif; Bolivia. Hutchings, a neuropsychologist, was a skilled technical climber who had led expeditions in Alaska and the Cascade Range. He knew about altitude sickness, he knew about ropes, and he was completely at ease on rock and in snow. The couple had considered climbing farther east, in Nepal, but had set their sights instead on the Zanskar Mountains in the Indian state of Jammu and Kashmir. For centuries, British colonists and Indian royalty had traveled to the region to escape the summer heat, and over the past twenty years it had become a mecca for Western trekkers who didn't want to test themselves in the higher areas of the Himalayas. It is a staggeringly beautiful land of pine forests and glaciers and—since an Indian government massacre of thirty or forty protesters in its capital, Srinagar, in 1990—simmering civil war.

The conflict had decimated tourism, but by 1995 Indian officials in Delhi had begun reassuring Westerners that the high country and parts of Srinagar were safe, so in June of that year Schelly and Hutchings headed up there with only the vaguest misgivings. Even the State Department, which issues warnings about dangerous places (and had Kashmir on the list at the time), will admit that Americans visiting such places are far more likely to die in a car accident than as a result of a terrorist attack. The couple hired two native guides and two ponymen (and their horses) and trekked up into the Zanskar Mountains. After ten days, on July 4, they were camped in the Lidder Valley, at eight thousand feet.

The militants, heads wrapped in scarves, secured Schelly and Hutchings's camp, rounded up a Japanese man and a pair of Swiss women who were camped nearby, and then left all of them under guard while the rest of the band hiked farther up the Lidder. A mile and a quarter away was a large meadow—the Yellowstone of Kashmir, as Schelly put it—that was guaranteed to yield a bonanza of Western trekkers. Sure enough, the militants returned to the lower camp two hours later with a forty-two-year-old American named John Childs, his native guide, and two Englishmen, Keith Mangan and Paul Wells. Childs, separated with two daughters, was traveling without his family.

The leader of the militant group, Schelly would learn later, was Abdul Hamid Turki, a seasoned guerrilla who had fought the Russians in Afghanistan and was now a field commander for a Pakistan-based separatist group called Harkat-ul Ansar. He ordered all the hostages to sit down at the entrance to one tent. Childs, nervous, looked down at the ground, trying to avoid eye contact with anyone. He was already convinced that the guerrillas were going to kill him, and he was looking for a chance to escape. A cold rain started to fall, and Turki asked for all their passports. The documents were collected, the militants attempted to read the papers upside down, and then they declared that all the Western men would have to come with them to talk to their senior commander. That was a three-hour walk away, in the village of Aru; they would be detained overnight and released in the morning, said the militants. Schelly was to walk to the upper camp with one of her guides.

"After I left, the men [were told] to lie down and pull their jackets over their heads, and that if they looked up, they'd be shot," says Schelly, who learned these details later from Child's guide. "The [kidnappers] went through the tents, stealing stuff. And then they took the guys off. By ten o'clock I'd gone to the upper camp and come back down [with the wife and the girlfriend of the Englishmen], and we all piled into one tent because we were still scared. I was awake at four the next morning, and I just kept looking down the trail thinking they'd be coming anytime now. It was six-thirty, and then seven, and then nine; that's when the knot in my stomach started."

Finally, Childs's guide returned. He had a note with him that he had been instructed to give to "the American woman." It said, "For the American Government only," and it was a list of twenty-one people the militants wanted released from Indian prisons. The top three were Harkat-ul Ansar.

The kidnapped men walked most of the night. They weren't being taken to the "senior commander"—he didn't exist—they were just being led deep into the mountains. The deception reminded John Childs of the tactics the Nazis used to cajole people into the gas chambers, and made him all the more determined to escape. The men walked single file through dark forests of pines and then up past the tree line into the great alpine expanses of the Zanskar Range. It was wild, ungovernable country the Indian Army didn't even attempt to control, and Childs believed that there was no way anyone was going to save them or even find them. They were on their own.

"I was convinced [the militants] were going to shoot us, and so as soon as I heard someone chamber a round into one of those weapons, I was going to take off into the woods," says Childs. "At one point we crossed a stream—it was snowmelt season, and the mountain streams were absolutely raging torrents—and I considered jumping in and flushing down to the bottom, but it would have been instant death."

Childs kept his eyes and ears open and waited. A chemical engineer for an explosives company, he was used to solving problems. This was just another one: how to escape from sixteen men with machine guns. There were personalities, quirks, rifts among his captors he was sure he could exploit. He started lagging while he walked, seeing if he could stretch the line out a little bit; he started taking mental notes of the terrain; he started probing for weaknesses in the group. "Escape is a mental thing," he says. "Ninety percent is getting yourself prepared to take advantage of an opportunity or create an opportunity. I knew that given enough time, I'd get away."

Late that night they came upon a family of nomads at a cluster of three log huts. The head of the family stepped out into the darkness to give Turki a hug. Then the militants and hostages all squeezed into the huts and fell into an exhausted sleep. A few hours later, as soon as it was light, Childs sat up and peered through a chink in the wall: alpine barrens and rock, nothing more. Escaping through the forest would have been at least a possibility, but crossing a mile of open meadow would be suicide. He'd be cut down by gunfire in the first twenty steps.

After the hostages were given a quick meal of chapatis, rice, and a local yogurt dish called *lassi,* they lined up on the trail and started walking again. This would become their routine in the next several days: up at dawn, hike all day, sleep in nomads' huts at night. The militants bought—or took—whatever food they wanted from the nomads and never had to carry more than a blanket and their guns. They told the hostages that they had been trained in Pakistan, near the town of Gilgit, and had come across the border on foot. They'd been in the mountains for months together and were prepared to die for their cause. When Donald Hutchings tried to engage them in talk about their families, one of the militants just patted his gun and said, "This is my family."

India and Pakistan have fought three wars over Kashmir, and Turki's band was the latest permutation in the fifty-year conflict. Harkat-ul Ansar (HUA) is committed to overthrowing Indian (thus, Hindu) rule in Kashmir and absorbing the state into the Islamic Republic of Pakistan. Since 1990 the militancy, as the rebel movement is known, has been waging a sporadic guerrilla campaign against Indian authority with automatic rifles, hand grenades, and other small arms acquired from Pakistan. Turki called the group he commanded Al Faran, a reference to a mountain in Saudi Arabia near where the Islamic prophet Muhammad was born. The first time anyone had ever heard of Al Faran was on July 4, 1995, when they came walking down out of the mountains into Schelly and Hutchings's camp.

The militants led the hostages by day through snowfields and high passes, traveling north. Childs's impression was that they were simply marking time in the high country, avoiding the pony trails in the valleys, where they might run into Indian soldiers. As the day wore on, the militants became less worried about being caught, and their vigilance slackened a bit. Turki remained dour and implacable, but the younger ones warmed up to the hostages. They called Don Hutchings *chacha*, meaning "uncle," and practiced their high school English whenever they could. Far from being threatening or abusive, they did anything they could to keep the hostages healthy: bandaging their blisters, giving them the best food, making sure they were warm enough at night. Not only did the hostages represent possible freedom for twenty-one separatists rotting in Indian jails, but they were also the only protection Al Faran had from the Indian military. They were a commodity, and they were treated as such.

"It was like a Boy Scout troop with AK-47s," says Childs. "The youngest militant was sixteen or so, a Kashmiri kid who'd been recruited to the cause. He hadn't been issued a weapon yet because he hadn't been through training; he was educated and bright, and his English was good. Turki was dead serious, though, and I didn't let any kind of camaraderie fool me. If he told them to kill someone, these guys wouldn't hesitate for a second; they were too well trained."

By the second day Childs had noticed an interesting—and horrifying—dynamic. The hostages, all desperately scared, turned to one another for comfort and support. They talked about their families, their homes, and their fears. But they had also been thrown into a ruthless kind of competition. They knew that if the militants were forced to prove their intent, they would shoot one of the hostages. That much was clear, but who would it be? An American? A Brit? A weak hiker? A brave man? A coward?

Since the hostages didn't know the answer, they did the next best thing: They tried to make as little an impression on their captors as possible. They didn't complain; they didn't cry; they didn't do *anything* that might cause them to be noticed. They blended in as completely as possible and hoped that if the time came to kill people, they'd be invisible to the man with the gun.

Childs quickly realized that he was losing the competition not to stand out.

"I was in really rough shape. My boots weren't broken in, and I'd gotten blisters on my heels before I was captured," he says. "The days went on, and the skin was rubbed literally to the bone. By the fourth day I was having trouble keeping up." Childs believed that he and Hutchings, the U.S. citizens, were the most valuable hostages, "but you could burn one American and still have one left over."

From time to time the hostages discussed the possibility of trying to escape en masse, but the consensus was that they would be putting themselves at terrible risk. Keith Mangan, in particular, was convinced that the situation would resolve itself peacefully. "Look, these things usually end without any tragedies," he told the others at one point. Childs wasn't so sure. Not only did he believe that Turki would kill them without a second thought, but he felt singled out for the first execution. Adding to his misery, he had come down with a devastating case of dysentery. By the end of the first day he was stepping off the trail every hour or so to drop his pants. A militant always gave him a Lomotil pill and followed him, so after a while Childs started relieving himself in the middle of the trail, in front of everyone. Soon they were waving him away in disgust, and he thought, "This is going to be useful. I don't know how, but it will."

The next time a militant put a pill in his mouth, Childs didn't swallow. He waited until the man looked the other way; then he spit the medicine out onto the ground.

It took six hours for Jane Schelly to hike out to Pahalgam, a jumping-off point for people heading into Kashmir's high country. The entire trekking population of the valley—some sixty or seventy people—was by the end walking out with her, and when they arrived at the Pahalgam police station, utter pandemonium broke out. Schelly informed an officer that her husband had been abducted, and she was taken into a back room and interrogated. "I had to decide whether to give them the note or not," she says, "because it said, 'For the American Government only.' I looked over at my guide, and he nodded and I thought, 'If they're going to help, let's get this show on the road.' So I gave it to them, they copied down the names, and then I went to the UN post. That was at eight P.M.; they called the [American] embassy, and things were kicked into motion."

The United Nations has had a presence in Kashmir since 1949, after Britain formally relinquished control of its Indian colony and the subcontinent sank into ethnic chaos. The British government's last administrative act was to draw a border between the Muslim majority in Pakistan and the Hindu majority in India, and that sent six million people fleeing in one direction or the other. Hindu mobs attacked trains packed with Muslims trying to cross into Pakistan, and Muslims did the same thing to Hindus going in the other direction. Trains plowed across the border between Amritsar and Lahore with blood dripping from their doors.

While half a million people were being slaughtered, the semi-independent state of Kashmir was trying to decide whether to incorporate itself into India or into Pakistan. Kashmir was primarily Muslim, but it was ruled by a Hindu maharajah, and that inspired an army of Pakistani bandits to cross the border and try to take Srinagar in a lightning raid. They were slowed by their taste for pillage, however, which allowed Indian troops to rush into the area and defend the city. War broke out between India and Pakistan, and the nascent UN was finally forced to divide Kashmir and demilitarize the border. Fighting continued to flare up for the next forty years, and a surge in Pakistani-backed guerrilla activity again brought the two nations to the brink of war in 1990. This time the stakes were higher, though: India had hundreds of thousands of troops in Kashmir, and both nations reportedly had the capacity to deliver nuclear weapons. Diplomats defused that crisis, but American envoys in Delhi still considered Kashmir the world's most likely flash point for a nuclear war.

By the time Jane Schelly and Donald Hutchings showed up in Srinagar, as many as thirty thousand locals had been killed since 1992, and Kashmir had been turned into a virtual police state. The brutal tactics employed by the Indian Army had brought a certain amount of stability to the area—it was alleged, for example, that security forces machine-gunned every member of a household that had any association with the militants—but the war continued to rumble on in the hills. In 1994 two Brits had been kidnapped by militants and held in exchange for twenty or so HUA guerrillas serving time in Indian jails.

The Indian government had refused to bargain, and after seventeen days the militants relented and let the hostages go. They even gave their prisoners locally made wall clocks as souvenirs of the adventure.

Schelly spent her first night out of the mountains at the UN post, and the next day she moved to a secure Indian government compound. High-level British, German, and American embassy officials flew up on the afternoon flight from Delhi, and by July 7 a formidable diplomatic machine was in gear. Terrorism experts—unnamed in the press—were flown in from London, Bonn, and Washington, D.C. Negotiation and hostage release specialists were made available to the Indian authorities. Surveillance satellites reportedly tried to locate the militants on the ground, and the Delta Force, a branch of the U.S. Special Forces, was in the area being readied for possible deployment. Indian security forces began working their informants in the separatist movement, and Urdu-speaking agents started trying to maneuver between brutally simple parameters for negotiation: no ransom and no prisoner exchanges. Any concession to the guerrillas' demands, it was feared, would only encourage more kidnappings.

Still, there was some hope that Al Faran could be eased toward compromise. Communication was carried out by notes sent along an impenetrable network of local journalists, militants, and nomadic hill people. Hamstrung, on the one hand, by an Indian government that was not entirely displeased with a situation that made Pakistan look bad and, on the other, by a U.S. policy that forbade concessions to terrorists, negotiators found themselves with almost no wiggle room, as they say. The best they could do was relay messages to Al Faran that pointed out the immeasurable harm the kidnappings had done to the Kashmiri cause; the only way to regain credibility, said the negotiators, was to let the hostages go. To encourage this line of thought, the U.S. government left the negotiating to the Indians—whose country it was, after all—and started pulling strings elsewhere in the Islamic world. They persuaded a Saudi cleric to condemn the kidnappings as un-Islamic, and they tried to massage some of their contacts in Pakistan.

"Al Faran was clearly an HUA-affiliated group, and what we know about HUA is that it's not very hierarchical," says a U.S. government source who closely followed the incident. "It's not at all clear that Al Faran was even *interested* in communicating with [HUA] headquarters. If we'd had anything suggesting a tightly hierarchical organization, it would have been much easier to negotiate. And they had very poor, unsophisticated decision making. These were *not* people with a Plan B."

By the end of the third day, John Childs could barely walk, and the militants seemed to be heading deeper and deeper into the mountains. They were, in fact, just walking in circles, dodging Indian military. To keep himself from sinking into despair, Childs devoted every waking moment to planning his escape. He knew the militants' sole advantage was their incredible mobility; without that, it would be only a matter of time before they were discovered by an army patrol. Which meant that if any of the hostages escaped, the militants wouldn't be able to waste too much time searching; they'd have to look quickly and then get moving again.

"My first objective was to get fifty meters away from them," Childs says. "And then five hundred meters, and then five kilometers. I knew that every bit increased the area they had to search by the square of the distance. And I knew there was no way this guy Turki was going to scatter his crew all over creation looking for me. He couldn't afford to look for me for more than six hours, so if I could stay away from them for that long, my only problem would be not being seen by the nomads."

That meant hiding during the day and traveling at night, which strongly favored an escape after dark. That was fortunate, because Childs had one iron-clad reason for getting up over and over again during the night: Dysentery was still raging through his insides. The militants always posted a sentry after dark, but the hostages weren't tied up when they slept, so the sight of Childs getting up to relieve himself was by now routine.

In contrast with Childs, the other hostages seemed to be doing fairly well. The two Brits, Wells, twenty-four, and Mangan, thirty-four, were depressed but physically strong, and Hutchings was fully in his element. When Mangan came down with altitude sickness, Hutchings had him pressure-breathing and rest-stepping as he walked; when the group got lost in a whiteout, he took charge and told them which way to go. At one point Wells muttered how he would like to grab one of the hand grenades and blow all the militants away, but Hutchings was always personable and helpful. "It's a lot tougher to kill a smiling face," he said. Hutchings had years of psychological training; if anyone could manipulate the situation, he could.

It wasn't until the fourth day, as they were crossing yet another valley, that the militants made their first mistake: They visited a familiar place. It wasn't much, but it was all Childs had.

"We were in the valley that the pilgrims take to Amarnath Cave," Childs says. "And Don knew where we were; he'd been there before. He said, 'Okay, down the valley is Pahalgam, and up the valley is the cave.' "

Childs thought about that for the rest of the day. He wasn't going to be able to keep up with the group for long, and Turki wouldn't hesitate to have him shot. Not only would that free up the group, but it would also send a message to the authorities, who obviously hadn't given in to the militants yet. If he were going to escape, he'd have to do it soon.

"So, are we going to spend the night here?" Childs asked Turki that afternoon, as they were taking a break. He knew the answer, but he wanted to hear what Turki had to say. "No, too much danger," Turki replied, waving his arm down the valley: Indian military. *They wouldn't dare spend much time searching for someone,* in other words.

That night the militants made camp along the east branch of the Lidder River, sleeping in a cluster of stone huts generally used by pilgrims on their way to Amarnath Cave. Childs, rolled up in a horse blanket, lay on the dirt floor of a hut and considered his possible avenue of escape. The camp was at the mouth of two huge valleys that fed into the valley leading to Pahalgam, and Childs's plan was to escape by climbing *up,* in the opposite direction of what the militants would expect. He'd hide in the snowfields before dawn, stay until dark, and then start down toward Pahalgam. It was a three-day walk, he figured; he had no food, no bedding, and the valley was filled with nomads who might report his location to the militants. It was, at best, a long shot, but it was better than the odds he had now.

Then, exhausted by four days of forced marches, Childs fell asleep.

"There had been other opportunities to escape, but of course you never know if it's the right time," says Childs. "It's not a movie, where you know when it's going to end. You keep asking yourself, 'Is this the best time, or will there be a better time with less risk?' It took a huge effort to focus my thoughts and say, *'Okay, you've got to do this now. You've got to do it when you're tired and not feeling well.'* "

Childs woke up in the middle of the night. It was quiet except for the sound of people snoring and the crash of the river. The dysentery was rolling through his system, so he fumbled in the dark and grabbed his hiking boots—knocking over a metal grate in the process—and crept out of the hut. Ordinarily the sentry would greet him and escort him out of camp. But this time no one stirred—the sentry seemed to be asleep. Childs walked out of camp, relieved himself, and then stole back into bed, wondering what to do.

"You can be passive and not make a decision that may save your life," he says, "or you can accept death as a possibility. That was the crux of the whole thing."

Childs lay in bed for an hour, preparing himself, and then he got up again. He decided that if anyone stopped him, he'd just claim he was having another bout of dysentery. He thought about waking up the other hostages, but there didn't seem to be any way to do that quietly; the others also lacked his pretext for getting up. Childs stepped out of the hut and waited for someone to say something; silence. He edged out of the firelight into the darkness beyond the huts; more silence. There was always the possibility that someone was watching him surreptitiously—or even had a gun trained on him—but that was a chance he had to take. He stood motionless for a moment, frozen at the point of no return, and then he started to run.

"I thought I was in their cross hairs the whole time," he says. "It was like a dream where you run and run and you're not getting anywhere because your feet are bogging down. I kept expecting to hear a ruckus behind me, but I never saw any of them again."

Childs took off straight up a ridge between the two valleys. He was in his stocking feet, and all he had on was long underwear, Gore-Tex pants, a wool shirt, and a pair of pile pants wrapped around his head. He walked and ran as hard as he could until the ridge got too steep to climb without boots, and then he put them on and kept going. He knew the militants would wake up early for morning prayers, and he had to get as far away as possible before then. He hammered upward for the next three hours, and when dawn came, he crept into a cleft in a rock, drew in a few stones to conceal himself, and settled down to wait. As it got lighter, he noticed that anyone walking along the ridge would stumble right into him, so he violated his rule against traveling during the day and continued higher up. He was in the snow zone now, really rugged country; the next hiding spot he found seemed perfect, until it became apparent that he was resting on solid ice. He wound up moving to a small patch of moss on a hillside. There were glaciers and peaks all around him, and he was sure no one would follow him up that high; he was at least at twelve thousand feet.

By midmorning a drizzling sleet had started to fall, and Childs endured that for a few hours—resting on the moss, dozing from time to time—before starting out for Pahalgam. He was almost down at the bottom of one of the side canyons when he heard a helicopter. The sound of the rotors faded in and out, then seemed to head straight toward him. Since he hadn't heard any aircraft in five days, his first thought was that there must have been a negotiated release of the hostages and now he was stranded in the high mountains with no food and no way to call for help.

"I stood there kind of dumbfounded," he says, "and I started waving my pants around over my head. The pilot circled, and I could see there was a soldier in there; he had his gun pointed at me. I was a mess by that point. I hadn't bathed in five days and had mud smeared all over me and looked like a wild man of the mountains. The pilot landed on one skid, I ran up, and [a soldier] said, 'Are you German?' I said, 'No, I'm American. I just escaped from the militants.' He said, 'It's a miracle from God,' and hauled me on board."

The militants, as Childs had thought, had searched down valley when they realized he was gone. They didn't find him, but they stumbled across two other trekkers, Dirk Hasert of Germany and Hans Christian Ostro of Norway. They were subsequently reported missing and were, in fact, the people the helicopter crew had been searching for. Rebel sources in Srinagar say that Ostro was belligerent toward the militants from the start, telling them that what they were doing was cowardly and un-Islamic; they also claim he was armed with a knife and had tried to use it. That is impossible to verify, but suffice it to say that Ostro succeeded in sticking out in the group.

Childs was brought back to Srinagar in triumph and immediately debriefed in the presence of British and American embassy officials. It was the first of endless debriefings over the next several days. "I spent more time in captivity by the State Department than by the militants," he said later. Childs was then taken to a secure guest house, where he was introduced to Jane Schelly. For Schelly, the chance to talk to someone who'd seen her husband only hours earlier was a relief beyond words.

"Do they have enough food and drinking water?" she wanted to know. "Do they have enough clothing? Do they know that the women are okay?"

Everything Childs had to say about the hostages was positive: They'd suffered no abuse, and the situation seemed similar to the peacefully resolved kidnappings of a year earlier. The current hostage team—referred to as G-4 because the governments of four Western countries were involved—had no reason to believe that this case would be different. While Indian security kept up a steady dialogue with Al Faran, the G-4 team continued to pressure Pakistan to intervene with HUA. (Pakistani officials were stubbornly claiming that the incident had been staged by India to make them look bad.) A rescue was deemed to be too risky; even Indian Army patrols were warned away from areas where the militants might be. Everyone, including the hostages, was worried that a surprise encounter could erupt into a firefight.

Childs flew back to Delhi two days after his escape, speaking to reporters at the airport despite the efforts of officials to bundle him into an embassy car. A few days later, he stepped off an airplane at Connecticut's Bradley Field, and news crews taped him sweeping his two young daughters up into his arms. He'd gone from the mountains of Kashmir to Hartford in the space of a week, and it rattled him. "Had circumstances been a little different, I'd be dead," he says. "You expect to live out your normal life span, but it could be over in a second. At the time I thought I'd never see my kids again. Now every breath I take is something I didn't expect."

While Childs was facing the news cameras back home, Jane Schelly was still in Srinagar, working frantically for the release of her husband. "Please let Donald go," she sobbed at a press conference, holding on to Keith Mangan's wife, Julie. "In the name of God, please let our loved ones go." Al Faran responded by passing along a statement that said they had let Childs escape on purpose, but that they would resort to an "extreme step" if India didn't release the HUA rebels. They also sent a photograph of the five hostages sitting on pine boughs in a stone hut, their hands tied behind their backs, their eyes downcast. On an accompanying tape, Don Hutchings said, "Jane, I want you to know that I am okay. But I do not know whether I will die today or tomorrow. I appeal to the American and Indian governments for help."

The G-4 team decided, for security's sake, to move Schelly, the German woman, and the two English women back to the British embassy's guesthouse in Delhi. Negotiations remained deadlocked, and one week later some very bad news came in: The militants had supposedly run into an Indian Army patrol near Pahalgam, and two hostages had been wounded in the ensuing gunfight. The Indian government denied that the encounter had taken place, so Al Faran released some photos showing Hutchings lying on the floor of a house with his abdomen wrapped in bloody bandages. There was no blood on his pants, though, and he seemed to be refusing to look into the camera—refusing, perhaps, to cooperate with the deception. The consensus at the U.S. embassy was that the photos had been staged, an opinion Schelly shared.

On an audiotape sent with the photos, Hans Christian Ostro asked the Indian government to give in to Al Faran's demands, pointing out that it was tourist officials in Delhi who had misled him into thinking Kashmir was safe. "I even went to the leader of the tourist office in Srinagar, and he gave me his card and said that if there was anything, I could call him," Ostro said at the end of the tape. "Well, Mr. Naseer, I'm calling you now."

Another week passed, and still there was no breakthrough in the negotiations. Britain's Special Air Squadron and Germany's elite counterterrorism force, the GSG-9, had by now joined the U.S. Army's Delta Force in Kashmir, even though an Entebbe type of rescue operation was unlikely; the authorities had no idea where Al Faran was, and there were also delicate sovereignty issues to work out with India. The feeling among the G-4 negotiators was that, as with the previous kidnapping, Al Faran would eventually give in.

They didn't.

On August 14, 1995, "we were at the German ambassador's residence, having lunch with the other families," recalls Schelly. "And during the meal several embassy people were called out of the room, and then more people were called out, and I didn't think anything of it. The German ambassador was called out just prior to dessert. We were eating cherries jubilee, and the next time I looked over, his ice cream had melted all over the place. And then I began to wonder."

While the families retired to a sitting room for coffee, a group of embassy officials talked somberly in a corner. Eventually one of them came over and reported that the body of a Caucasian man had been found in the village of Seer, outside Srinagar, but they didn't know if he was one of the hostages.

In fact, they did know, but they weren't saying. Cars came to pick up the families, the Ostros' car arriving first. After they had pulled away, the German ambassador put his arm around Schelly and said, "It's not your husband." It was not until that moment that Jane Schelly finally accepted the possibility that she might never see her husband again.

The body was that of Hans Christian Ostro. The guerrillas had cut off his head, carved "Al Faran" in Urdu on his chest, and dumped his body by an irrigation ditch. His head was found forty yards into the underbrush, and a note in his pocket warned that the other hostages would suffer the same fate if the HUA prisoners weren't released within forty-eight hours. The families of the remaining hostages were told that Ostro's chest had been carved after he was dead, that he had been "peaceful" when he died, and that he hadn't been killed in front of the other hostages—though how the officials could know that is unfathomable. However peaceful Ostro's death, though, he may have known it was coming: Medical examiners found a good-bye note hidden in his underwear.

The G-4 team—now down to G-3—responded by demanding proof that the other hostages still lived. The militants passed along a photograph of the four holding a dated newspaper and also arranged for a radio conversation between Donald Hutchings and the Indian

authorities. At ten forty-five on the morning of August 21, a negotiator raised the guerrillas on a military radio, and Hutchings was put on:

"Don Hutchings, this is one-zero-eight. When you are ready,
please . . . tell me 'One, two, three, four, five.' "

"One, two, three, four, five."

"The first message is . . . from your families. Quote, 'We are all
staying together in Delhi and we all send our love and prayers. We are
helping each other. Be as strong as we are.' Over."

"Okay, I have the message."

"Now, Don Hutchings, there [is] a set of questions for you. You'll
have to provide me with the answers because I don't know them. . . .
Am I clear to you?"

"Yes."

"What are the names of your pets? I repeat, what are the names
of your pets?"

"My pets' names are Bodie, B-O-D-I-E, and Homer."

Hutchings's existence was confirmed. The negotiator continued
with personal questions for each of the other hostages and then signed
off, telling Hutchings to "have faith in God and strength in yourself."
Within days of the radio interview, Al Faran began renewing their
threats to kill the hostages, and their tone was so antagonistic that
members of G-3 privately admitted that they thought the odds of the
hostages' surviving were only fifty-fifty. September crawled by, and
then October, and the winter snows started to come to Kashmir.
Reports of frostbite and illness among the hostages began to drift in.
And then, on December 4, the inevitable happened: Al Faran ran into
the Indian Army.

The guerrillas were passing through the village of Mominabad
early in the morning when a patrol of a dozen Indian soldiers spotted
them from the marketplace and someone opened fire. According to
Indian military officials, there were no hostages with them—they were
presumably being held nearby—but that's impossible to confirm. The
militants jumped a barbed-wire fence, splashed across a shallow

stream, and then ran through a patch of scrub willow. They headed across a dry rice paddy, machine-gun fire hammering behind them, the villagers diving into their mud houses and slamming their doors shut. The militants made it across the paddy and took a stand farther upriver, near the small village of Dubrin, and the Indian patrol called for reinforcements. Soon dozens of troops were firing on the rebels, who held off the army for six hours until dark fell, when they left their dead and ran.

Turki was killed, along with four other Al Faranis. Three days after the gunfight the British ambassador in Delhi received a phone call from a man claiming to be with Al Faran and offering new terms of release: $1.2 million in ransom and safe passage to Pakistan. "You know, you know, we have been treating [the hostages] as our guests for the last five months plus," he complained. "You can expect that we have spent lots of money." The ambassador demanded proof that the hostages were still alive, but the man never called again.

And that was it. From time to time, over the next few months, nomads reported seeing the hostages up in the mountains, but those reports came to be suspect when it was revealed that the nomads were making money both as paid police informants and as messengers and suppliers for the kidnappers. In April 1996 a captured HUA militant claimed that the hostages had been executed about a week after the fight at Dubrin, in retaliation for Turki's death, and that the bodies were buried in a village called Magam. The Indian Army scoured the woods and fields around Magam for weeks without finding anything.

"You want to be optimistic, your heart says be optimistic, but your mind says, 'Sucker, you've gotten your hopes up before,' " says Schelly. "We had a full moon right before Id-ul-Fitr [a feast day at the end of Ramadan], and a friend of Don's said, 'This is the last full moon that's going to pass before Don comes back.' I was so convinced they would release him for Id-ul-Fitr that I packed my bags and got my hair cut. At one point I had to pull the car over on the way home from work and throw up, I was so worked up."

Id-ul-Fitr came and went, as did the one-year anniversary of the kidnappings, without any word from Al Faran. Reports continued to trickle in from the nomads, but nothing could be confirmed. Schelly went back to Kashmir in the summer of 1996 to meet with HUA leaders, and she returned there a few months later to start up a reward program. Announcements in local newspapers, on local radio shows, and even on the backs of matchbooks offered money to anyone who would come forward with information. The U.S. government also offered a reward, and India followed suit.

"It's very difficult to say if the program will be successful," says Len Scensny of the State Department's South Asia bureau. "We haven't had verifiable contact with the hostages in over a year, and we have no current information on their well-being. It's been an ongoing subject of discussion with very senior officials in both India and Pakistan."

Meanwhile, John Childs has resumed his life in America—working, jogging, spending time with his daughters. People who know Childs have made joking references to Rambo, which bothers him, and some even ask why he didn't help the others escape. It's a question that still troubles him. "I rationalize it and say, 'I couldn't have done it any other way,' but without having done it another way I'll never know," he says. "I ask myself constantly, 'Should I have done anything different?' Sitting here in my office it's one thing, but when I actually made the decision to escape, I was tired, I was injured, I was miserable, I was terrified. It revealed something about my character, and I'm not even sure if I'm proud of it or not."

And Jane Schelly's hopes are slowly waning. While promoting the reward program in the fall of 1996, she decided to visit the village of Seer, where Ostro's headless body had been found. She talked with the villagers through an interpreter and then walked along a dirt path by the irrigation ditch where, among the rice paddies, two women had spotted the body a year earlier. "It was so incongruous," says Schelly.

"The village was on a little pass, and when I was there, everyone was harvesting the rice. There [were] stacks of rice straw in the fields and mountain peaks in the distance. It was probably one of the most beautiful spots I've ever seen in my life."

If Don Hutchings is still alive, he's probably looking out at a scene very much like that one: iron gray mountains, a scattering of mud huts, and a dozen villagers cutting their way across the rice paddies at dusk. One of them, undoubtedly, knows Hutchings is there; one of them, undoubtedly, wonders if telling the army would put his family in jeopardy. He decides to say nothing. And Don Hutchings, peering out through a chink in the wall, watches night come sweeping up his valley one more time.

KOSOVO'S VALLEY
OF DEATH
1998

It wasn't much of a town, Prekaz, just a dozen or so farmhouses strung along a dirt road that ran between some low brown hills. In the distance were the mountains of Albania, and all around were the dead winter fields of Kosovo. The houses had red tile roofs, thick whitewashed walls, and traditional courtyards—a defensive layout that probably hadn't changed much in the past eight hundred years. The pastures began at the road and stretched up to the crests of the hills before ending in ugly swatches of scrub oak. It was the kind of scrub oak that would whip you in the face if you tried to run through it. It was the kind of scrub oak that you could disappear into.

Before dawn on March 5, 1998, hundreds of Serb special police took up positions on the hilltops around Prekaz. There were mortar emplacements, tanks, heavy artillery, 20-mm cannon, and dozens of armored personnel carriers mounted with heavy machine guns. It was the first premeditated military assault by a European government against its own citizens since Nicolae Ceauşescu unleashed his Romanian security police in 1989—and that was basically the last spasm of a dying government. Before that you'd have to go back to the Nazis. Kosovo, a province of Serbia, is only two hundred miles from Italy; tourists come to ski in the winter. There hadn't been a war here since 1945. And now one of its towns was about to get scraped off the map.

The attack started with an artillery barrage against one household and quickly escalated to a ground assault against the entire village. Police in greasepaint and black uniforms poured out of armored cars and moved down the sodden brown hills, firing automatic weapons and rocket-propelled grenades. Mortar shells dropped into the houses and lit them on fire. Albanian separatist guerrillas were said to be holed up in the town, and the Serbs weren't taking any chances: They weren't going to let the bastards surrender, and they weren't going to let them hide. If necessary, everyone would die.

Women and children took shelter until they realized it was only a matter of time before they were killed, and then they took their chances and ran through the gunfire into the woods. The men weren't so lucky. Some fought back, and others just hid; either way, they died. They died as their houses collapsed on them; they died as automatic-weapons fire ripped through the cinder-block walls; they died on their doorsteps as they tried to surrender.

"The soldiers shouted for us to come out one by one or they would kill us," the daughter of a man named Šerif Jašari later told human rights workers. "When my cousin Ćazim came out with his hands up, they killed him on the steps. We ran and had just gone through the first cordon when the soldiers caught my cousin Nazmi, who was helping his mother, Bahtije, along. They grabbed him, tore off the woman's dress we had given him to wear, and ordered him to lie down on the ground and then to get up. He had to do this many times. Then they fired into his head and back, and I saw his body jerking from the bullets."

The next person the Serbs shot was the girl's seventeen-year-old brother, Riad, hitting him twice. He fell to the ground, and his sister and mother took him by the arms and started dragging him into the woods. "We went through the second cordon posted in the street outside the house. Armed soldiers in green uniforms with yellow and black markings and the same colors smeared on their faces," Jašari's daughter said. "We hid in the bushes, and up on the hill we met some people we knew, and they drove my brother Riad to a safe place. When Bećir's wife, Sala, arrived, she said they had shot Bećir in the leg and that he had told her to go with the children. A few days later we heard Bećir was dead."

Bećir Jašari was a member of a wealthy Albanian family that was said to be involved in an Albanian independence movement in Kosovo. Kosovo is about 90 percent ethnic Albanian but remains part of the Serb-dominated former Republic of Yugoslavia, which stripped it of its autonomy in 1989.

Tensions in the area had been rising steadily since November, when three masked Albanian guerrillas appeared at the funeral of a man killed in a crossfire between Serb police and guerrillas. "The Kosovo Liberation Army is the only force which is fighting for the liberation and national unity of Kosovo!" one of them shouted, and the mourners—twenty thousand strong—responded, "U-Ç-K!" the Albanian initials for the Kosovo Liberation Army. The opposition movement in Kosovo was headed by a longtime pacifist named Ibrahim Rugova, but it also had an armed wing ready to take the fight into the hills.

Almost immediately after their appearance at the funeral, the KLA began ambushing police cars and sniping at the checkpoints. Then a car chase and shoot-out in late February resulted in the deaths of four policemen and five KLA members. Another badly wounded KLA fighter reportedly dragged himself to the home of Ahmet Ahmeti in a nearby village called Likošane. Like the Jašaris, the Ahmetis were a wealthy family rumored to have links to the KLA.

On February 28 the Serbs struck back. Attack helicopters blasted towns with gun and rocket fire, and policemen in black uniforms dragged people out of their houses and shot them on their doorsteps. Twenty-six were killed. Witnesses said the Ahmeti men over the age of fifteen were separated from the women and children, savagely beaten, and then executed in their courtyard with shotgun blasts to their heads. One had his eyeballs dug out. Journalists who later visited the house reported that the ground was littered with teeth and hair and that a human jawbone hung from a nearby bush.

There was a brief lull while people buried their dead, and then the police moved in on Prekaz, which lay only a few hundred yards from an old munitions factory that had been converted into a barracks for the Serb special police. On the morning of March 5 the police stepped outside their front gate and attacked. Some snipers didn't even bother leaving the compound. Fifty-five people died in Prekaz, including thirty from the Jašari family alone.

One of the few Jašaris who survived was an eleven-year-old girl named Besarte, who had hidden under a heavy slab on which her mother used to make bread. She remembers shells crashing into the house for hours and her uncle Adem singing folk songs "so the family wouldn't lose its faith in life." When the bombardment finally stopped, the bodies of her entire family lay twisted around her. Twenty-four hours later—after another night of siege—several policemen stormed into the house to check for survivors. One stopped in front of Besarte, who played dead, but he put his hand to her chest and felt a heartbeat, so he picked her up and took her to the munitions factory. She arrived spattered with blood, screaming that she wanted to stay with her sisters.

I arrived in Kosovo two weeks after the massacre, on a frigid March night. I drove in with an old friend named Harald Doornbos, a Dutch journalist who had been based in Sarajevo since 1992. For obvious reasons, the Serbs weren't granting entrance visas to journalists, but Herald knew a dirt road border crossing into Montenegro where the guards—being Montenegrin—couldn't have cared less what the Serbs wanted. From Montenegro we could easily cross into Kosovo.

We got up early the next morning to try to drive into Drenica, the rural stronghold of the KLA. We crossed a desolate brown plain and plunged into the hill country, the little towns flicking by in our car windows and the mountains on the Albanian border looming in the distance. Guns were coming in over those mountains; Albania was awash in weapons, and the KLA was completely dependent upon help from across the border. The Serb military reportedly had shoot-on-sight orders for anyone in the high peaks, and soldiers regularly ambushed Albanians moving weapons into Kosovo over mountain tracks.

There were said to be KLA training camps inside Albania; in response, the Serbs have massed a tremendous number of heavy weapons at the Albanian border—far more than are needed to stop arms smuggling. The fear is that the Serb Army will cross into Albania to stamp out the camps and that the situation will escalate into an all-out war between Yugoslavia and Albania. Such a conflagration could drag in Greece and Turkey and—in a worst-case scenario—divide the United Nations. Another scenario has it that war in Kosovo might trigger a similar war in Macedonia, which has its own restive Albanian population, and that Greece and Bulgaria could jump in to grab Macedonian land that they have old claims to. More than three hundred American troops are stationed in Macedonia to contain exactly that kind of domino effect, but they are scheduled to be withdrawn this summer.

The towns we passed were dead and empty-looking, and house after house stood half built, abandoned by Albanians who could no longer afford to finish them, because they'd lost their jobs in Serb-controlled businesses. After half an hour we turned down a dirt road and drove until we dead-ended at a railroad tunnel near a river. We stopped, grabbed our notebooks, and walked through the tunnel and into an empty brown valley surrounded by brush-covered hills.

We were worried about KLA snipers—stupidly, both of us were dressed in black, like the Serb secret police—but we were even more worried about Serb snipers. This was the heart of Drenica, an area the police can seal off but not control, an area the KLA can hide in but not defend. It was a no-man's-land where you could get shot at or you could get invited in for tea, depending on who spotted you first.

We walked for an hour and finally came upon a dozen ethnic Albanians repairing the road. Since the Serb police controlled the highways, there seemed to be a lot of repair work being done on the spider web of dirt roads that connect the villages in Drenica. The men escorted us into one of their houses and sent someone ahead to ask the KLA commander at the next village if we could continue. We sat on the floor, drinking Turkish coffee and watching an American cop show on satellite television; after an hour the man came back and said apologetically that the answer was no, we could not continue. The KLA was not prepared to greet us.

When we stepped outside, we could hear the Serbs shelling some villages a few miles away. The sound rumbled over the hills like a summer rainstorm. As we studied the faces of the farmers around us—rough, unshaven faces of men who had known nothing but hard work their whole lives—it was impossible to tell if they understood what real war would mean. It was impossible to tell if they understood that tragedies like this happen every day, all over the world; that in all probability, no one was going to intervene on their behalf; and that the Serb authorities, like most governments, would stop at almost nothing to retain power.

In 1389, as the myth goes, Prince Lazar of Serbia was visited by St. Ilija in the form of a falcon. It was on the eve of a great battle with the Turks, and Lazar had gathered around him, on the plains of Kosovo, much of the Balkan military elite: Bosnian warlords, Albanian noblemen, and Hungarian horsemen with shamanic bones sewn onto their uniforms. Lazar was understandably nervous—the Turks had wiped out an entire Serb army eighteen years earlier—and wondered if it might not be better to retreat and fight again another day. St. Ilija gave Lazar the choice between a kingdom on earth and a kingdom in heaven; Lazar, wisely choosing the kingdom in heaven, went on to meet his death at the hands of the Turks.

The battle became known as the Battle of Kosovo Polje—the Blackbird Field—and it occupies a particularly fevered part of the Serb psyche. It was on Kosovo Polje that a Serb leader first chose death over subjugation; it was on Kosovo Polje that the guiding maxim of the Serb people, "Only unity saves the Serbs," was first acted out in all its bloody glory.

Nearly six hundred years after the battle, Slobodan Milošević, the man responsible for igniting the entire Balkan conflict, would stand on the ancient battlefield and whip a crowd of angry Serbs into a nationalist frenzy. "Yugoslavia does not exist without Kosovo!" he yelled, instantly catapulting himself to the top of the political heap. "Yugoslavia would disintegrate without Kosovo!"

Kosovo is not the birthplace of the Serb people, however. The original Serbs migrated southward from Saxony and what is now the Czech Republic in the sixth century A.D. and didn't settle permanently in Kosovo for another six hundred years. The high-water mark of the Serb empire came in the 1330s, when a brutal nobleman named Stefan Dušan defeated his own father in battle, had him strangled, and then went on to extend his empire throughout Kosovo and into Greece. He built numerous Orthodox monasteries and churches and eventually had himself crowned emperor of the Greeks, Bulgarians, Serbs, and Albanians.

The empire didn't survive his own death, though; within decades the Turks defeated the Serbs at Kosovo Polje, and three hundred years after that the Turks put down another uprising so ruthlessly that most Serbs fled Kosovo. The void they left behind was filled by the Albanians, who drifted back down out of the mountains with their wild, hill people ways.

Traditional Albanian society was based on a clan system and was further divided into brotherhoods and *bajraks.* The *bajrak* system identified a local leader, called a *bajrakar,* who could be counted on to provide a certain number of men for military duty. In another era Adem Jašari and Ahmet Ahmeti might well have been considered *bajrakars.* That organization has fallen into disuse, but the clans— basically used to determine allegiances during a blood feud—seem to have survived.

Feuds in this part of the world inevitably break out over offenses to a man's honor, which include calling him a liar, insulting his female relatives, violating his hospitality, or stealing his weapons. Tradition dictates that these transgressions be avenged by killing any man in the offender's family, which creates another round of violence. As late as the end of the nineteenth century, one in five adult male deaths was the result of a blood feud, and in Albania today, it is said, a tradition still exists whereby you must kill one man for every bullet in the body of your dead kin.

Seen in the context of the code of male honor, the Serb police have violated just about every blood feud rule in existence, including the killing of women—a provocation above all others. It's no wonder they have such a hard time maintaining control of Kosovo.

The Kosovars were granted autonomy at the end of World War II, but then aspiring President Milošević had the autonomy revoked in 1989, and the Dayton Accords of 1995, which ended the recent war in Bosnia and Croatia, failed to address the issue of Kosovo's status. Inevitably, an independence movement was born, funded by a voluntary 3 percent income tax given by the Albanian diaspora and supported by groups in Albania proper.

The first armed clashes in Kosovo were reported during the summer of 1995, and within two years the KLA was strong enough to force a column of Serb armored vehicles to retreat from Drenica. After that the Serbs began a slow buildup of police and heavy weapons in Kosovo and on the Albanian border, culminating in the attack on Prekaz.

If anything, the massacres have radicalized the youth of Kosovo. The Serbs have already spent an estimated six billion dollars controlling the province. In some ways, they couldn't have engineered a worse domestic problem if they'd tried; in some ways, they fell right into the KLA trap.

The next morning dawned cold and gray, with a mean little wind blowing trash down the streets; the cafés in town were completely empty. We packed the car and drove out of the city by a different route, hoping to drive into Drenica over some dirt roads that skirted the Serb checkpoints outside Prekaz. We wanted to see the villages that were getting shelled. The Serb government had bowed to international pressure and agreed to resolve the dispute through diplomacy, but meanwhile it was hammering the villages with rocket and artillery fire.

We had no problems at the first checkpoint—just the usual guns in our faces. But at the second one a police officer in an army jumpsuit stormed over and ordered us out of the car. He was young, cleanshaven, and handsome in the way that Serb men often are: black hair, light skin, pale blue eyes. "You journalists are all spies!" he screamed at Harald. "You always make Serbs look bad! If I had my way, I'd tear the skin right off your faces!" He ripped the passports out of Harald's hands and studied them while unloading a steady stream of hate. The guards were all standing around us with their machine guns leveled at our bellies. Finally the head cop came over and handed my passport back to me. "We know where you live," he said darkly. "Write the truth or we'll find you and kill you."

As checkpoints go, it could have been worse—far worse. Albanian translators have been arrested and beaten at checkpoints, and the day before the attack on Prekaz, Harald and three other journalists were punched, dragged into a bunker, and questioned for an hour. When the police saw that Harald lived in Sarajevo, they accused him of being a Muslim—the predominant Albanian religion—and Harald had to prove he wasn't by making the sign of the cross. Then the cops started going through Harald's notebooks, demanding a translation of every word that was written down.

At one point, a cop spotted the name Frenki Simatović in Harald's notebook, then turned to his friend and said, "Look, he even has the name of our boss in here." Harald had no idea who Simatović was; he'd just written the name down during an interview and filed it away for future reference. Then they demanded to know if any of the reporters had ever been to a town called Prekaz. They kept asking over and over again: "Prekaz? Prekaz? Have any of you motherfuckers ever been to Prekaz?"

Prekaz is such a small town that before the massacre, people in Priština—a city half an hour away—had never heard of it. Harald just kept pleading ignorance, but when the Serbs finally released him, he called his editors and told them to be on the lookout. "I have no idea where it is; it's not on the maps," he said. "But something's about to happen there. Just check the wires for a town called Prekaz."

The next morning the first shells started to fall.

Back in 1991, as Yugoslavia began its descent into the hell of civil war, the newly elected Milošević had a somewhat delicate problem on his hands. He wanted to drive the Croats and Muslims out of large swaths of Yugoslavia, but he didn't dare send the Yugoslav Army to do it.

The solution he came up with was simple. First, he surrounded himself with a trio of rabid nationalists—Jovica Stanišić, Radovan Stojičić, and Frano ("Frenki") Simatović—known collectively as the Vojna Linija, or the Military Line. The Vojna Linija had little association with the Serb Army; it was a shadowy group within the Ministry of Interior Affairs, which was known as the MUP. After the Vojna Linija was established, Milošević began arming local Serb populations in Croatia and Bosnia, and training paramilitary forces. The weapons, distributed by Stojičić and Simatović, were taken from police and army depots. The paramilitary forces simply came out of the country's jails.

According to Marko Nicović, a former Belgrade police chief who later had a falling-out with Milošević, convicts were told that their sentences would be suspended if they went to the front lines. Many were only too happy to oblige. The best-known groups were the White Eagles of Vojislav Šešelj, a virulent conservative later named to the Belgrade government; the Red Berets of Frenki Simatović; the unnamed forces of Captain Dragan; and—worst of all—the Tigers of Željko Ražnatović. Arkan, as Ražnatović was known, was wanted by Interpol for bank robberies and murders committed throughout Europe.

In 1992 the Yugoslav Army officially withdrew from Bosnia, but Serb paramilitary forces, including Simatović's Red Berets, continued to operate there. That same year Šešelj and Arkan went to Kosovo to terrorize the locals into peacefulness, opening a recruiting office in Priština's Grand Hotel and putting snipers up on the rooftops. (They also made tremendous amounts of money on the local black market.)

Both men turned up around Srebrenica in 1993, "cleansing" the Muslims from the small towns in eastern Bosnia. The Dayton Accords left the paramilitary foot soldiers without much to do, so they either sank back into Belgrade's underworld or looked for other wars; some reportedly fought—and died—in the jungles of Zaire during the downfall of Mobutu Sese Seko. They didn't have to wait long for another war in their own country, though: by 1997 Kosovo had ignited.

Harald and I had been in Kosovo about a week when things started to calm down; we could almost joke with the police at the checkpoints. The Serbs were still shelling the villages in central Drenica, though, and before leaving Kosovo, we decided to make one more stab at going there. We went in on a big, sunny day, the shadows of cumulus clouds sweeping across the Drenica hills and the fields mottled and bare in the early-spring sunlight. We were headed for Ačarevo, a town rumored to be the center of KLA resistance.

There were two ways to get in: walk six miles along some railroad tracks and hope no one shot at you, or drive down dirt roads across the central plateau and hope no one shot at you. The cops at the checkpoint warned us that there was a lot of gunfire on the road and suggested that we wear flak jackets. We thanked them and drove on, and as soon as we were out of sight we turned onto a dirt track that we thought led to Ačarevo.

The road climbed up onto a plateau, and we started across the highlands of Drenica, like some huge, slow beetle scratching across someone's dinner table. "I don't like this," Harald muttered. I rolled down the window so we could hear gunfire more easily, and soon the landscape of war magically materialized all around us: bunkers and machine-gun nests and tanks on distant ridgetops. They emerged out of nowhere, like images brought out by a darkroom developer. But when I looked away, it took me a moment to find them again. They were there; then they weren't. "This is crazy," Harald said. "The entire fucking Serb Army is watching us."

He turned the car around, and we plunged back down the dirt road and went jouncing out onto the hardtop. It was difficult to see how the KLA could fight a guerrilla war in a land like this: no forests to hide in; no mountains to run to; no swamps to stop the tanks. Just open fields and brush-choked hills. It would be suicide to confront the Serbs openly on such ground, so the KLA's only choice is to carry on a war of harassment that may eventually cost the Serbs so much, in money and lives, that they have to pull out.

For their part, the Serbs have no stomach for a protracted fight in which farm kids from Drenica are popping out of the hedgerows with grenade launchers and AK-47s. A grenade launcher will easily take out a tank; a Molotov cocktail placed in its air intake will destroy one as well. The Serb population—largely spared the horrors of Bosnia but demoralized by massive inflation and a crippled economy—isn't going to stand for a war in which too many of its young men get roasted alive in their tanks.

For the Serb military, the only solution is terror. Every time a cop is killed, wipe out a family. Every time a police patrol gets shot up, level a village. Slaughter is a lot easier—and cheaper—than war, and it forces the young idealists in the KLA to decide whether they really want this or not. It's nothing for a twenty-four-year-old with no future and no civil rights to sacrifice his life in a guerrilla movement; it happens all the time. But for him to sacrifice his kid brother and two sisters and mother: that's another question entirely.

Harald and I continued north on a small paved road until we topped out on another hill, from which, far away, we finally saw Ačarevo. It wasn't much, just a small white village shoved down between some hills. It rippled in the heat coming off the fields. We moved on, and around the next bend we found ourselves at a heavily reinforced checkpoint, with mortars by the road and bunkers dug into the hillsides. We stopped, and a cop came out cradling a machine gun. "Let me see your papers," he said. He stood there studying them for a while as I sat sleepily in the passenger seat and Harald lit up a cigarette.

The sniper must have been waiting for a car to pass so the cop would have to step out into the road. He must have been lying there in the scrub oak, smoking cigarette after cigarette, completely wired with this new killing game, contemplating how he was going to escape when he finally lost his nerve and stopped shooting. The place was crawling with Serbs; he'd have only a few minutes to get out of there.

The first shot simply caused the cop and me to look at each other in puzzlement. The second one got Harald and me out of our seats. The third forced all of us—me, Harald, the cop—to dive behind the car. It's amazing how fast animosity vanishes among people who are suddenly getting shot at. One cop fumbled with his radio; the others shoved their guns over the tops of the sandbags as they tried to figure out where to return fire. *Pap . . . pap . . . pap.* The guy on the radio shouted for help while Harald and I scrambled across the road and into the bunker. The cop next to us struggled to put on his flak jacket with the resigned look of someone who had to do this at least once a day.

The shooting stopped as suddenly as it had begun, and a cop dismissed us with a wave of the hand. "Get the fuck out of here," he told us. We got back into the car and drove out of the highlands, past a town called Lauša—shot to pieces in the offensive—past the Serb police headquarters in Srbica, and right up to the gate of the munitions factory. The dirt road to Prekaz crosses in front of the gate, and we drove down it slowly, not wanting to give the impression that we were trying to slip past anyone.

The paramilitary soldiers didn't stop us until we were right on the edge of town, coming at us out of a camouflaged bunker, with guns drawn and incredulous expressions on their faces, as if they couldn't believe someone was stupid enough to defy them. They looked as if they would have stopped even a regular police car; they looked completely uncontrolled by anyone but themselves. One of them shouted for our papers while two others circled the car, guns trained on us. "We were just shot at by the KLA," Harald said out the window. "Now we understand why you guys are here."

It worked. The soldier studied our papers and then waved us through. As far as I could tell, the only reason the Serb military allowed journalists into Prekaz—a damning place, easily sealed off—was to spread word of what would happen to those who resisted.

Harald drove slowly down the town's wide dirt street, which ended at a pasture. A dead cow lay rotting by the side of the road. Every house had its roof blown off, its windows shot out, or its walls caved in. Rooms spilled their contents to the world, as if disemboweled by some huge claw. Walls were pocked with mortar shell explosions;

tongues of soot licked roofward out of windows. Bullet shells lay in gleaming little piles wherever someone had really put up a fight.

Harald and I walked through a wooden gate, splintered by artillery, and into the courtyard of a house. Two abandoned dogs, one with a wound on its back, growled at us from what used to be the doorstep of their home. Harald gave the dogs some sausage and a tin of sardines, and we stepped around them and into their family's home. Schoolwork sat on tables, and jackets hung on pegs alongside things that had been blown to bits. It was odd what had been touched and what hadn't.

After the attack this particular house had served as an outpost for the special police, who had gone through it room by room, laying their hands on everything that could be tipped over or broken open. Books, clothes, photo albums, and lamps all lay tangled on the floor. On top of one pile was a Serb porn magazine, discarded by the latest occupants.

We paid our respects to the fifty-five rectangles of freshly dug-up earth in the pasture above town, and then we drove back out to the world of the living. As we passed, the men at the bunker were posing for a group portrait—the destroyed town in the background, their machine guns wedged upright in the crooks of their arms. The men grinned broadly at us.

One of them wasn't holding a gun in his hands. He was holding a huge double-bladed ax.

DISPATCHES FROM
A DEAD WAR

1999

EDITOR'S NOTE: In July 1974 the Turkish military seized the northern third of Cyprus after a violent coup by right-wing Greek Cypriots—supported by Greece—appeared to threaten the Turkish Cypriot minority. Twenty-five years later Cyprus remains partitioned, a case study in how ethnic hatred perpetuates itself, but perhaps also a manual on how peace might be sustained in places like Kosovo. In February Harper's Magazine *sent Scott Anderson and Sebastian Junger to report on this intractable zone of conflict. To decide who would go to which side, they flipped an old Greek coin with a man's head on one side and a war chariot on the other. The coin landed chariot side up, which meant that Anderson traveled to the Turkish Republic of Northern Cyprus (TRNC) and Junger to the Greek side, the Republic of Cyprus.*

Sebastian Junger
REPUBLIC OF CYPRUS

> *A fool throws a stone into the sea*
> *and a hundred wise men cannot pull it out.*
> —CYPRIOT PROVERB

The rusting yellow car sits on four flat tires against an old wall in the buffer zone, directly in front of a cement bunker with a machine-gun slit. Painted a cartoonish camouflage, the bunker is manned by a lone Greek Cypriot soldier, who smokes a cigarette as he watches us.

I have been walking the length of the buffer zone in Nicosia with a British peacekeeping soldier named Murphy, who carries a silver-tipped walking stick instead of a gun. He uses it to point things out. We've started at a UN observation post at the east end of town and progressed between the two irregularly parallel cease-fire lines under a dreary rain that patters through the thick no-man's-land foliage and fills puddles in the road. Murphy has shown me where, in 1989, a Greek Cypriot soldier supposedly dropped his pants and from 164 feet away mooned his Turkish counterpart, who promptly shot him dead. The spot, now a patrol landmark, is identified by a sign: MONUMENT TO THE MOON. Farther along is a place where the UN-patrolled zone, known in Nicosia as the Green Line, squeezes down to the width of a narrow street. The balconies of two buildings on either side extend to within ten feet of each other, and a few years ago Greek and Turkish soldiers took to strapping knives to the ends of long poles and jousting with each other. In other places they sling stones or shout insults.

"We can't do anything about it unless we see it happen," Murphy tells me. "It's all right for the [Greeks] to say, 'These Turkish soldiers are throwing stones at us.' . . . So we phone up the Turks and say, 'We've had reports that some of your soldiers are throwing stones.' The first thing they say is, 'Well, did you see it?' And we say, 'No, we didn't.' So there's not a lot we can do."

Now we stand in the rain in front of the old yellow car, which also is identified by a sign, YELLOW CAR. A landmark for UN patrols, the car was once the focus of a bitter dispute between the Greeks and the Turks. In the original delineations of the buffer zone, Turkish territory was described as extending from the "front" of the yellow car to the corner of a building. By "front" the UN meant the end of the car where the headlights are located. The Turks, however, argued that the front was the end of the car nearest to one of their observation posts; the resulting difference in the angle of the cease-fire line would give them another fifty square feet of territory.

"They finally worked out a compromise," Murphy tells me. "The line stayed where it was, but a Turkish soldier gets to stand in the triangle of disputed territory for five minutes each hour."

The Green Line was established in 1963 by a British commander who was trying to quell street fighting that had erupted between Greek and Turkish militias. He supposedly took a green pencil and bisected a map of Nicosia from one side of the old Venetian fortifications to the other. Eleven years later, after the Turkish Army overran a third of Cyprus, the buffer zone was extended across the length of the island, a distance of 112 miles. A few months later the United Nations Peacekeeping Forces in Cyprus [UNFICYP] oversaw a massive, but peaceful, population transfer of 40,000 Turkish Cypriots from the south to the north to replace the estimated 175,000 displaced Greek Cypriots, most of whom had fled south during the invasion. The exodus was proclaimed voluntary as well as temporary, but of course it was neither. When the Turkish Republic of Northern Cyprus finally declared itself to be an independent state in 1983, all but the most optimistic refugees realized that they were never going home.

Today the two countries mark their borders as the cease-fire lines that were established in 1974. Between the lines is the buffer zone that none of the 190,000 Turkish Cypriots to the north or the 655,000 Greek Cypriots to the south may enter without special permission. Per capita, Cyprus is the most militarized country in the world after North and South Korea—with 35,000 Turkish and Turkish Cypriot troops and 14,500 Greek Cypriot troops, monitored by 1,200 UN soldiers— yet it is one of the most peaceful: only 16 people have been killed along the divide since 1974. Greek Cypriots refer to the buffer as the dead zone. On Greek Cypriot maps, the Turkish Republic of Northern Cyprus is labeled "Area Occupied by Turkish Troops," and in conver-

sation, Greek Cypriots often refer to it as the so-called Turkish
Republic or simply the pseudo-state. There are no embassies or con-
sulates in the TRNC besides Turkey's, and the UN does not maintain
formal diplomatic relations with them. There is a checkpoint at Ledra
Palace, in the middle of the buffer zone on the western edge of
Nicosia, but only foreign passport holders may cross through it, and
only from the south to the north and then back again. You cannot go
in the other direction, and if you visit the TRNC, you must be out by
5:00 P.M.

Within Nicosia the Green Line doesn't look like much, just a series of deserted streets that end at brick walls and cement barriers.
Every so often appears a sandbag bunker with a Greek Cypriot soldier
inside, invariably smoking a cigarette. The line has a strange pull to it,
like the edge of a cliff or a third rail; it was the first place I went when
I arrived in Nicosia. I dropped my bags at the hotel and walked past
the fancy shops on Ledra Street to a cul-de-sac, where some staging
had been set up against a concrete wall along the line. It's the only
place where tourists can look out over the rubble of no-man's-land,
and a flight of metal stairs has been installed to encourage viewing.
While I was there, an English family arrived and trudged dutifully up
to the platform, children licking at ice-cream cones and parents fid-
dling with camcorders. They looked over the railing at the ramshackle
Turkish positions a hundred feet away, clucked their disapproval, and
had their photo taken with a young soldier who was standing guard
nearby. Then they wandered off to do more shopping.

The soldier had an M-16 slung around his neck and spoke fair
English. I asked him if he and his buddies ever talked with the Turkish
soldiers on the other side, but he told me that this was the one spot
on the Green Line where the Turks don't post guards. Apparently,
tourists who step up to the platform occasionally get carried away
and start yelling, and the Turks don't want to deal with that.
Elsewhere, though, the Turks will shout insults at the Greeks or throw
rocks.

"Do you ever yell back?" I asked the Greek soldier.

"No," he said, smiling. "We are careful not to provoke them, because we are the weaker side."

It was a strange admission for a soldier to make, though in keeping with the general theme of the lookout point. Alongside were a photo exhibit of the wartime destruction and a map showing, day by day, the changing battle lines of the Turkish invasion. Few countries would offer up such evidence of their own worst defeat; it was practically a monument to Turkish military might. The point seemed to be that Cyprus was the object of unbridled aggression from a highly militarized government and that if the world didn't act decisively, who knew what would happen next?

Thirty years ago it was the Turkish Cypriots who had to be careful not to provoke. The problems started in earnest in late 1954, when two Greek gun-running boats made the 250-mile crossing from the island of Rhodes to Cyprus and landed on a deserted stretch of the western coast. On board were hundreds of pounds of explosives and a former Nazi collaborator named General George Grivas, who had arrived to lead a guerrilla group called the National Organization of Cypriot Fighters. Known by its acronym, EOKA, the group was committed to kicking the British out of Cyprus—they'd been there since the Ottomans handed it to them in 1878—and eventually uniting Cyprus with mainland Greece. The prospect of union with Greece—"enosis"—presented a terrifying threat to the 18 percent Turkish minority in Cyprus, however, who in no way wanted to become Greek citizens. So it was with considerable alarm that they watched three hundred EOKA guerrillas, fighting with pipe bombs and homemade machine guns, elude twenty thousand British troops and forty-five hundred Cypriot police in the rugged Troodos Mountains. By 1959 the British still hadn't been able to stamp out EOKA, so they gave the Cypriots their independence—and thus made Cyprus the rest of the world's problem.

It was clear to the West that given the level of rhetoric, General

Grivas wasn't going to stop until he had achieved union with Greece, an outcome that Turkey would never permit. The south coast of Turkey lies only forty miles away, and a Greek military presence so close to its borders was unthinkable. If the enosis movement were to succeed, Turkey would invade Cyprus, Greece would intervene, and suddenly there would be—at the height of the Cold War—a full-blown conflict between two NATO members.

To prevent such a disaster, the British arranged for a meeting in Zurich between the antagonists. They finally agreed to a fabulously awkward constitution that provided for a Greek Cypriot president, a Turkish Cypriot vice-president, and disproportionately large Turkish representation in the parliament. England was to retain two military bases on the island, and both Greece and Turkey were allowed to con-tribute small contingents of troops for common defense. As signato-ries to the agreement, England, Greece, and Turkey all could intervene militarily if they deemed the Cypriot constitution to be in danger. On August 16, 1960, the Republic of Cyprus was born, with a former EOKA leader, Archbishop Makarios III, as president. Almost from the beginning the arrangement was a nightmare.

It was the contention of the Greek Cypriots that the Turkish Cypriot minority had no reason to fear for their safety and that hatred between the two groups was the result of Turkish propaganda and British manipulation. ("As late as 1955 Greeks and Turks had always lived peacefully together, like brothers," reads a placard at Nicosia's Museum of National Struggle. "Their relations had always been com-pletely harmonious, and the Turks had never put forward any claim on the island.") In reality, things had never been so rosy. Although they had tolerated each other for centuries, Greek and Turkish Cypriots had largely lived in separate communities, and calls for enosis drove the two groups even farther apart. By the early 1960s death squads of Greek na-tionalists were regularly killing Turkish Cypriots, who, instead of turn-ing to the government for protection, started to gather into easily defended enclaves and arm themselves. In retaliation, the Greek

Cypriots tried to strangle the Turkish communities with economic blockades, and the situation quickly escalated into gun battles in the streets. By late 1963 the Green Line had been established across Nicosia, but even that didn't stop the fighting, and Archbishop Makarios finally appealed to the UN for help. Several thousand peacekeepers were sent in with a renewable ninety-day mandate, but by then the Turkish Cypriots had completely severed relations with the Cyprus government, and fighting was breaking out regularly between the two militias.

Like a bad marriage, the split was only a matter of time. In the late 1960s Archbishop Makarios officially stopped calling for enosis as a political goal, and in July 1974 he accused the Greek military of trying to undermine his power. A cadre of right-wing officers, outraged by what they perceived to be a betrayal of Hellenism, sacked the presidential palace and chased Makarios into hiding. They also killed hundreds of moderate Greek Cypriots suspected of being Communist sympathizers or simply soft on Turks. Within days they had replaced Makarios with an EOKA fighter named Nikos Sampson, who had already proved his patriotism by taking seven hundred Turkish Cypriot civilians hostage during the Green Line clashes ten years earlier. Within forty-eight hours the Nixon administration had dispatched a high-level diplomat named Joseph Sisco to try to keep Turkey out of the war, but it was already too late. "We will not repeat the mistake we made ten years ago," the Turkish prime minister told Sisco on July 19. The next morning a flotilla of Turkish troop carriers scraped ashore near the north Cyprus town of Kyrenia and disgorged six thousand Turkish troops.

Scott Anderson

THE TURKISH REPUBLIC OF NORTHERN CYPRUS

I will tell you a story about Cyprus. Once there was a snake, and one day this snake came into the house of a man who had a son.

The snake bit the man's son and that son died, so in his grief the
man took up a knife and cut off the snake's tail. The next day the
snake came back and said to the man, "Okay, now let's be friends."
The man said, "We can never be friends, because you killed my son,
and that is a pain I will carry in my heart forever, and I cut off
your tail, and that is a pain you will carry in your heart forever."
So that is why there can never be peace in Cyprus.
—ELDERLY TURKISH CYPRIOT WOMAN

An old man and a scruffy white dog stand at the edge of an empty
swimming pool, both seemingly lost in thought as they stare into its
depth. The pool is exceptionally deep—maybe fifteen feet—and lined
with cracks, its bottom covered with a thick layer of dead leaves. The
man spots me on the opposite side of the gate and beckons me
through.

"Very bad design," the man mutters when I come alongside. "Big
problems."

I ask if he's thinking of repairing it.

"No, no." He chuckles. "It has been like this for twenty-five years.
It is a museum." He looks to the three-story house beyond; it is an an-
gular structure, concrete balconies and windows perched above the
sea. "All this is a museum. In 1974 it was the home of [President]
Makarios's doctor; now it is for the Peace Operation martyrs."

In the early-morning hours of July 20, 1974, advance units of the
Turkish amphibious force started coming ashore in a small cove about
three miles west of this house on the north coast of Cyprus. It marked
the beginning of what Turkish Cypriots still call the Peace Operation.
A matter of definition, perhaps, because the most immediate results of
that operation were the deaths of as many as four thousand soldiers and
civilians, the dislocation of over two hundred thousand more, and an
international crisis that very nearly led to regional war. I'm not here to
quibble, though; the old man starts toward the house, and I follow.

It was a cold overcast day, and I had headed west out of the coastal resort town of Kyrenia to explore the nearby beaches where the Turkish soldiers had first come ashore in 1974. I had stopped at a memorial park on a bluff overlooking the sea, an austere mausoleum with the graves of some seventy Turkish soldiers arrayed before an abstract sculpture of bent black metal. To one side lay another kind of graveyard, some two dozen old tanks and armored personnel carriers parked in neat rows and surrounded by flower beds and trees. Most of the weaponry appeared to be of 1950s vintage, the feeble armor the Greek Cypriots had mustered to oppose the Turkish Army, and the joint ravages of combat and pilferage had transformed them into empty husks. It was while walking amid the tanks that I had glanced over the fence to see the man and his dog by the swimming pool.

At the entranceway to the house, the old man stops and draws my attention to the scars in the flagstone wall. "This is where they killed Karaoglanoglu," he says, referring to the Turkish ground forces' commander killed early in the invasion. He points to a nearby clump of trees. "The Greeks were hiding in there, and when Karaoglanoglu peacefully approached this door—*tok!*—a mortar." He shakes his head sadly, then pushes open the door and motions me inside.

The far side of the house is a wall of windows, and just beyond is the Mediterranean, all whitecaps and thrashing waves on this stormy day. The bottom floor is taken up with display cases of captured Greek weapons, fragments of shells and grenades. Upstairs are four rooms, each lined with row upon row of black-and-white photographs of young men in formal pose, Turkish soldiers killed in the Peace Operation. Some of the photos appear to be from high school graduations, the teenagers in civilian dress and smiling, whereas others look to be enlargements of military identification cards, the subjects more somber and with shaved heads. Here and there are glass display cases containing the dress uniforms of dead officers and their personal effects: wallet-size photographs of wives or children or girlfriends, letters home written on thin paper, medals.

If not much of a Peace Operation, the first phase of the Turks' 1974 invasion was also not much of a military triumph. In fact, it was pretty much a fiasco, a detail glossed over by the Turkish government but given unintended confirmation by the neat juxtaposition of the rows of "martyr" photographs in the oceanfront museum with the display of antique enemy weaponry in the adjacent field.

On that first day, all had gone rather smoothly for the Turkish soldiers. Coming ashore at the western end of Five-Mile Beach, the six-thousand-man vanguard had met little resistance and by evening had fanned out along the coastline; in the morning, commanders planned to cross over the Kyrenia Mountains and link up with the paratroop unit that had landed just outside Nicosia—or Lefkosa, as it is known to the Turks. It was with nightfall that things began to go awry.

Incredibly, the naval armada that had delivered the vanguard to the Cyprus coast headed back to Turkey at dusk, and so did the jet fighters that had provided air cover throughout the day. Even more incredibly, there was virtually no communication link between the landing force on the island and military planners on the mainland. Greek Cypriot fighters, complemented by Greek soldiers, seized the moment to attack all along the Turkish line, surrounding the paratroop unit outside Lefkosa and streaming down from caves in the mountains above Five-Mile Beach to fall upon the landing units strung out along the coast. Throughout the night, ferocious battles raged as positions were overrun, retaken, and lost again in a chaos of close combat made worse by raging brushfires.

At dawn the Turkish Air Force finally returned to the skies, and what had been a seesaw battle now turned into a slaughter. Turkish planes bombed military positions across the island, decimated Greek Cypriot armored convoys caught in the open, and cleared their entrenched mountain positions with napalm. By the time a cease-fire was declared the next day, the Turkish Army had carved out a narrow enclave that extended all the way to the Turkish Cypriot neighborhoods of Lefkosa.

But the Peace Operation wasn't done just yet. Over the next three weeks, as diplomats frantically sought a solution to the crisis, Turkey quietly built up its force on Cyprus to some thirty thousand troops, and they were ready to roll when the peace talks collapsed. In just three days the Turks poured out of their bridgehead to seize more than a third of the island and create the frontier they still hold today.

It's all a little hard to picture at ground level, however. Up close, the Turkish Republic of Northern Cyprus resembles nothing so much as a quiet, slightly raffish tourist destination. The once-pretty villages along the northern coast have been transformed into sprawls of cheap hotels and fish restaurants, weird concoctions in faux-Tudor or -Bavarian style to lure the British and German vacationers who predominate. Those who tire of lolling on the beach can take hikes in the hills, visit the ruins of castles, or play the slots at one of the grim roadside casinos. Along with tourism, the TRNC has become an international tax haven—"an ideal place for foreign businessmen," government brochures exhort—with an array of dubious-looking offshore banks tucked away in the back streets of Lefkosa.

Lurking at the edges of the landscape, however, is a parallel universe: the martyrs. Over the past twenty-five years the Turkish Cypriots and their mainland Turkey protectors have studded the countryside with monuments and cemeteries and museums dedicated to those who have fallen, and the message these buildings and fields carry is directed equally at the villagers in the hills and the tourists on the beach: This is a land created by blood and defended by blood; there can be no return to the old days.

In the story of their existence, the Turkish Cypriots weave an epic tale of victimization and dominance. From their vantage point, history has been a four-hundred-year-long siege in which the majority Greek Cypriots have never ceased trying, through both force of arms and guile, to force them into an intolerable union with Greece—or to push them off the island altogether. Nowhere does this mythology more radically diverge from that of the Greeks than in the interwar period of 1964 to 1974, between the collapse of the republic and the arrival of Turkish troops.

In the Greeks' telling, this was the island's golden age, an idyllic time when the two communities coexisted in harmony. In the Turks' rendition, it was the time when the noose was steadily tightening around their necks, when they were forced to seek safety in tiny vulnerable enclaves, and any trip outside the "ghettos" meant constant harassment by Greek Cypriot authorities or worse. With the bloody EOKA coup against Makarios in July 1974, Turkish Cypriots figured that they were the next targets for annihilation, once the Greek moderates were dealt with, making the Turkish Peace Operation a justified act of defense.

That sentiment is firmly on display in the monument built above the little cove on Five-Mile Beach where the Turkish soldiers came ashore. Just down from a great pillar of concrete jutting out of the ground at such an angle as to be nicknamed the Turkish erection are seven concrete stele that purport to tell, in brief words and bad etchings, the history of modern Cyprus.

The first two stele borrow heavily from Picasso's *Guernica:* lots of unhappy people and animals afloat in flames. By the third panel, help is on the way: Lantern-jawed Turkish soldiers stride into the fiery wasteland with drawn swords, their progress heralded by flittering doves of peace. For the rest of the monument, the warriors for peace continue apace, the flames gradually tamped out, the doves joined by blooming flowers and pretty—if slightly lantern-jawed—girls.

Other honorifics to the Turkish Cypriots' version of a martyr-filled history are scattered throughout the TRNC. The former Greek fishing village of Ayios Yeoryios, where Colonel Karaoglanoglu was killed, has been renamed in his honor, and Five-Mile Beach is now officially the Beach of the Resolute Outbreak. Beside the old Venetian wall of Famagusta is a little graveyard with a sign in Turkish, English, and German that reads: ARMED GREEK CYPRIOTS AND GREEK THUGS TRIED TO ELIMINATE EVERYTHING TURKISH TO ACHIEVE *ENOSIS;* IN THIS

CEMETERY LIE TURKS WHO, UNARMED AND DEFENCELESS, WERE MAR-TYRED BY GREEK CYPRIOTS AND GREEKS. In Lefkosa the government has built a Museum of National Struggle, perhaps to maintain parity with the Museum of National Struggle on the Greek side of the city.

In at least one sphere of the museum competition, however, the Turkish Cypriots have achieved hegemony. Located on Mehmet Akif Boulevard, the Museum of Barbarism is a single-story whitewashed building set in an unruly yard of flowers and fruit trees and surrounded by much newer and taller structures: chrome and glass auto dealerships; five-story office buildings. The particular acts of Greek barbarism it is dedicated to are those of 1963, and the curators have clearly opted for the scared-straight approach. Lining the walls of the foyer are a dozen ghastly black-and-white photographs of Turkish Cypriots of all ages lying dead in fields, in morgues, being exhumed from burial pits, their bodies bullet-riddled, knife-slashed, decomposed.

The Museum of Barbarism stays on its theme. In each of its several small side rooms, each lit with a bare light bulb dangling from the ceiling, is an unbroken line of similarly grotesque photographs mounted at eye level. Some of the captions identify the victims and detail the circumstances of death, but others are more general: "Another innocent victim of the brutal Greek campaign to exterminate the Turkish population."

In the museum's largest room, one comes to an ominous display, a glass sarcophagus filled with bath towels and baby shoes. Instead of murder photos, two walls of this room are lined with the personal snapshots of a young family. One shows a young boy at a table crowded with other young boys, staring at a large cake set before him. "Murat, pensive on his seventh birthday," the caption reads. "With his left hand on his cheek, he tries to guess what the future has in store for him on this happy occasion, not knowing of course that he has few days left to live."

As it turns out, in 1963 this little house on Mehmet Akif Boulevard was the home of an army doctor, Major Nihat Ilhan; his wife, Muruvet; and their three young sons, including seven-year-old Murat. The major happened to be away when EOKA gunmen attacked on the night of December 24, but they found his family huddled for safety in the bathtub; the next morning, a photographer dutifully recorded the grisly scene. That photograph—Muruvet Ilhan lying dead in the bathtub, her three dead boys clutched to her chest, the bathroom walls and floor sprayed with blood—is now such an iconic image in Turkish Cyprus that the museum curators have hung several large copies of it, along with two paintings that seek to replicate the scene faithfully, as if only repetition can convey its awfulness.

But it is more than just an image, for beyond the sarcophagus of bath towels is the bathroom itself, untouched for thirty-six years. Beneath a coat of dust, a white bottle of liquid soap still stands on the edge of the bathroom sink, and the tub and walls bear the same cracks and bullet holes as in the photograph. "The marks on the ceiling," a small sign above the tub reads, "are brain pieces and blood spots belonging to the murdered."

Sebastian Junger
REPUBLIC OF CYPRUS

Posted at the Ledra Street viewpoint—alongside photos of Greek refugees and buildings pancaked by Turkish bombs—is a list of what the Turks gained by invading Cyprus. According to the Cyprus government, the Turks gained 70 percent of the island's gross output, 65 percent of the tourist accommodations, 83 percent of the general cargo capacity, and 48 percent of the agricultural exports. Those are just numbers, though; Greek Cypriots generally don't grab you in bars and complain about their loss of cargo capacity. They grab you and complain about the city of Famagusta.

Famagusta lies near the center of a long scallop of bay on the eastern shore of the island, facing Syria. In the thirteenth century it was the wealthiest port in the Mediterranean, and before 1974 its Varosha district, now a Turkish military base, was the most fashionable beach resort on the islands, flooded every spring by English and Scandinavian tourists. Like Nicosia, it is surrounded by massive stone walls that were reinforced in the sixteenth century by Venetian military engineers bracing for the arrival of the Ottomans. The invasion finally came in 1570. As any Greek Cypriot can recount, fifty thousand Turks came ashore in the withering heat of midsummer, led by a sadist named Lala Mustafa. After sacking Nicosia and killing twenty thousand of its inhabitants, Mustafa led his forces against Famagusta, which was defended by a garrison of Venetian soldiers.

The Turks hammered the thick stone walls with an estimated hundred thousand cannonballs until the Venetian commander, Marcantonio Bragadino, finally ran out of supplies. Bragadino arranged for peaceful terms of surrender, but the Turks, enraged by the losses they had suffered while taking the city, started torturing and killing Bragadino's soldiers. When Bragadino objected, Mustafa ordered that his ears and nose be cut off and that he be skinned alive. The skin was stuffed with straw and mounted on a wagon, and legend has it that Bragadino lived long enough to behold his own gruesome double paraded through the streets of Famagusta with a parasol stuck in its arms.

Four hundred years later the Turkish Army walked back into the city. It was August 1974, "phase two" of the Turkish invasion, and Famagusta's Greek Cypriot inhabitants had grabbed whatever they could and fled south to the small farming town of Dherinia. From a gently sloping hill they could look down on the beautiful beaches and now-empty hotels that had been their home just hours before. It almost certainly didn't occur to them that the situation was permanent; it almost certainly didn't occur to them—late in the twentieth century—that upper-middle-class Europeans could be driven out of their beach homes by a modern army without the rest of the world intervening. They were wrong.

Now the closest they can get to Famagusta is a hillside several miles away where they can look out at the city. Two "viewpoint" cafés have sprung up, each boasting Turkish-atrocity photos and a rooftop viewing platform. I stop at the one called Annita's because it also over-looks a spot where two Greek Cypriots were killed by TRNC forces in 1996. Annita's is a three-story apartment building on the edge of a desolate swath of Dherinia suburb that abuts the buffer zone. Across the street are a roll of razor wire and then several hundred yards of un-tended fields and then more razor wire. A flagpole flying the TRNC flag—a red sickle moon and star against a white background—marks the beginning of "the pseudo-state."

I climb three flights of stairs to the café, sit down at a table, and order a coffee. It arrives with a pair of binoculars. On the wall of the café is a stop-action sequence of a young Greek Cypriot named Tassos Isaak getting beaten to death in a field; the field is the one I can see out the window. Next to the photos is a placard: "On the 11th of August, 1996, the barbarian Turkish settlers brutally murdered in cold blood and in full view of the UNFICYP, Austrian contingent, a peace-loving 24-year-old Cypriot. They used truncheons and metal bars to crush the spirit of freedom."

The events that led to Isaak's death were set in motion when the European Federation of Motorcyclists organized a ride to protest the Turkish occupation of Cyprus. One hundred and twenty riders left the Brandenburg Gate in Berlin on August 2, 1996, and proceeded on a one-week tour of Europe. They wound up in Cyprus on August 10, and, after joining forces with some seven thousand bikers from the Cyprus Motorcycle Federation, promptly declared their intention to crash the cease-fire line. After pressure from the UN, Cypriot President Glafcos Clerides finally forced the bikers to change their plans, but thousands of protesters gathered in the Dherinia area anyway. The Cyprus police were deployed along the cease-fire line near what is now Annita's but had left the checkpoint unmanned, and by midafternoon the protesters had pushed their way into the buffer zone and started screaming at the Turkish troops. They were quickly confronted by a rough crowd of a thousand Turkish Cypriots who had been bused in by the Turkish military. The Turkish counterdemonstrators were predominantly civilians but carried bats and iron bars, and some were members of a vicious nationalistic group called the Grey Wolves, who had come from Turkey to deliver—in their words—"a special surprise package" to the motorcyclists.

Watching all this was Rauf Denktash, president of the TRNC, recording the events with a camera and telephoto lens. A melee broke out in the buffer zone, and as Turkish troops started firing into the crowd, four Greek Cypriots—including Isaak—got hopelessly tangled up in razor wire. UNFICYP policemen managed to pull three of them free, but Isaak fell to the ground and was quickly surrounded by an ugly knot of Grey Wolves. Photographs taken from the Greek Cypriot side show him desperately trying to ward off the blows while Grey Wolves and Turkish police officers in riot gear take turns beating him on the head with truncheons and iron bars. By the time UNFICYP peacekeepers managed to get to him, Isaak was dead.

A wall-mounted television at Annita's café plays, in a continuous loop, news footage of Isaak's death, as well as footage of the next death three days later. On the afternoon of August 14, immediately after Isaak's funeral, a few hundred motorcyclists returned to the same spot outside Annita's and again managed to get past the Greek Cypriot police into the buffer zone. Among them was Isaak's cousin, twenty-six-year-old Solomos Solomou. Footage of the second protest shows Solomou dodging past two UNFICYP soldiers and slipping through a gap in the fence that separates the buffer zone from Turkish territory. Waiting for him were a line of Turkish troops, machine guns at the ready, and a cluster of state security officers on the balcony of a nearby building. Solomou managed to cross the Turkish cease-fire line and make it to a large white pole that was flying the TRNC flag. While security officers leveled their weapons at him and UNFICYP soldiers looked on in amazement, Solomou started shinnying up the pole.

He made it about a quarter of the way up before a red splotch blossomed on his neck and he slid back down to the ground. A total of five bullets hit him in the stomach, neck, and face. News photographs clearly show two security officers—later identified by the Greek Cypriot police as Kenan Akin, now a TRNC member of parliament, and Erdal Emanet, chief of the TRNC special forces—firing pistols from the building, quickly followed by Turkish troops kneeling and firing into the crowd of protesters. Two UNFICYP soldiers and seven Greek Cypriots were wounded, including a fifty-nine-year-old woman who had shown up to try to convince her son to come home.

I scan the buffer zone with the binoculars that came with my coffee, but it just looks like every other weeded-over field I've ever seen. The windows of the two-story building that the Turkish security forces fired from have been bricked in, with slits left for machine-gun barrels, and the TRNC flag still flies on the pole that Solomou tried to climb. I watch the video loop of the killings several times and then get back in my car and drive around until I find the cemetery where Isaak and Solomou are buried. It's a small plot of stone crypts surrounded by a concrete wall, tucked behind the town's soccer stadium. Isaak's grave is crowded with flowers, and several plastic-coated photographs of his own murder are propped against the gravestone. Solomou's gravestone is fancier. It depicts, in poured concrete, Solomou on the flagpole as Turkish soldiers level their guns to kill him. In a war with few casualties, along a front line with almost no gunfire, his tomb serves to remind people that there's still an enemy out there.

"The tragedy of Cyprus is that there is no tragedy," goes a sarcastic bit of local wisdom. The idea that there hasn't been enough suffering to merit world intervention is blasphemy, of course, but there are still a few Greek Cypriots who believe this. They just have to be quiet about it. Later, after returning to Nicosia, I ask a longtime European diplomat what he thinks of the idea.

"Both sides revel in this sort of victimology," the diplomat says, asking not to be identified. "It's what we call a double-minority problem, where both sides feel like they're the oppressed minority. The Turkish Cypriots say that their security is threatened because they are a minority on the island. The Greek Cypriots argue that they're a minority if you take Turkey and Cyprus together. . . . And neither side will stand up to its obligations as an equal player in this dispute, so both sides wait for the other to take the first step."

The diplomat works in an ultra-high-security office near the Ledra Palace checkpoint. Out his window I can see a huge Turkish Cypriot flag marked out in stones on a distant hillside. Turkish troops supposedly went up there day after day and painted the design on the undersides of the stones. When they were done, they waited until nightfall and then turned all the rocks over. The next morning, the Greek Cypriots awoke to find a huge Turkish Cypriot flag emblazoned across the flanks of the Kyrenia Range.

"Is there a solution?"

"The problem could be solved if you had cooperation between Greece and Turkey," says the diplomat. "Which is not on the horizon. If you look at Northern Ireland—I don't like drawing parallels, but this is quite a good one, actually—up until 1984 Britain and Ireland were at loggerheads, and the communities in Northern Ireland exploited this difference to ensure that the conflict just raged on. Then the British and Irish governments agreed to a joint policy on Northern Ireland and stuck to it, firmly. The two communities could not see any light between the policies of the two governments, and in the end they just had to come to terms with each other. If you had that kind of cooperation between the motherlands, the Cyprus problem could be solved pretty easily."

At the end of the interview the diplomat takes me up to the roof for a look at Nicosia. The sun is setting behind the Troodos

Mountains, and we can hear the Muslim call to prayer drifting over from the north side of town. The buffer zone runs like an awkward scar through it all, and beyond it are the massive earthen berms of the Turkish defenses, dug in with tanks and artillery. The diplomat points out the slapdash Greek defenses on our side and then traces the course of the buffer zone as it extends west. "It's filled with songbirds and wild animals," he says. "Hunters have killed everything else on the island, and it's the one place they can't go."

Scott Anderson
THE TURKISH REPUBLIC OF NORTHERN CYPRUS

Rauf Denktash, the president of the Turkish Republic of Northern Cyprus, doesn't much look the part. A short, portly man of seventy-five who bears a striking resemblance to Homer Simpson, he speaks English with just a trace of a British inflection—a result of his legal training in London in the 1940s—and is most often photographed in baggy sweat suits. On this day, sitting in his office in the heavily guarded Presidential Compound in downtown Lefkosa, he wears a business suit. The office is spacious and sunlit, and he shares it with a large aquarium of tropical fish and three very noisy parakeets, in a cage beside his massive desk.

For over four decades Denktash has been the dominant political figure in the Turkish community of Cyprus. One of the chief organizers of the outlawed Turkish Defense Organization back in the 1950s—and twice expelled from Cyprus for his violent militancy—he has been president of the TRNC since its founding. Obviously, such a man knows how to parry journalists, and the evening before our meeting I'd asked a local reporter the best way to handle him.

"Above all, don't ask him anything historical," the journalist urged. "As soon as you give him the chance to mention the constitution of

1960, you're doomed; you're going to get the Denktash history lesson for the next half hour."

Well, forewarned is forearmed. Sitting across from the president at the couch and coffee table arrangement in one corner of his office, I ask my first, carefully designed question.

"I'm already quite familiar with the history of Cyprus and because I know you're a very busy man, I'd like to concentrate on what is happening today, on what you feel is most important for Americans to know about the TRNC and the current situation in Cyprus."

The president nods. "What I would like Americans to know is that Cyprus has two owners, Greek Cypriots and Turkish Cypriots, and these two owners had agreed to form a partnership republic in 1960."

As the journalist suggested the previous night, "doomed." With never a pause, Denktash begins his discourse on the island's modern history from the Turkish viewpoint: the rise of the EOKA terrorists in the 1950s; the 1960 London Agreement, which the Greeks immediately sabotaged; the terror that existed in the Turkish enclaves throughout the 1960s; how the 1974 Turkish Peace Operation undoubtedly saved them all from EOKA annihilation; the political stasis that has existed ever since.

"And what do you see as the ultimate solution to the Cyprus problem?" I finally manage, because even the most energetic seventy-five-year-old has to pause sometime.

"A bicommunal confederation," Denktash says. "That's it. The Greek Cypriots must recognize our legitimacy and our right to govern ourselves. We've never made any claim on them—we've never called Cyprus a Turkish island, we have always recognized that we share this small island with them—and they must view and treat us the same way. I have said this to the Greek Cypriots many times, and they have always refused to hear it."

Underlying Denktash's comments is a deep resentment of the Republic of Cyprus's ability to keep his domain isolated from the rest of the world. Since Turkey is the only country in the world that officially recognizes the TRNC, it means that international flights do not land there, all diplomatic missions are kept at the "interests section" level, and all incoming mail is routed through a drop box on the Turkish mainland. On the flip side, the isolation gives offshore companies in the TRNC an enormous advantage over companies that have to adhere to international standards and helps fortify Denktash's state of siege message to his people.

In the Greek Cypriot worldview, Rauf Denktash is either the consummate political opportunist, his power dependent on his ability to keep the island divided, or a puppet of mainland Turkey and its "occupation" forces. In reality, Denktash appears to be enormously popular across the political and social strata of the TRNC. With a repetition that is at first quaint, then becomes tedious, his countrymen have the habit of calling him "the father of our nation" and make frequent comparisons to Kemal Atatürk, the founder of modern Turkey. At times it seems that almost everyone in the country, whether expatriates along the north coast or farmers in the most remote and impoverished mountain village, has had some surprise personal encounter with the president. Usually these involve Denktash, a serious photography hobbyist, tramping through the countryside in his baggy sweat suit with a camera around his neck, his small security detail following at a discreet distance. And although there certainly are those who feel that he is getting too old for the post, his political power hasn't diminished; in each of the five presidential elections he has stood for, Denktash has emerged triumphant.

Even more remarkable is the degree to which his take on the "Cyprus problem" and how to resolve it is shared by his countrymen. If a visitor to the TRNC is not careful, he or she will be subjected to the "Denktash history lesson" by virtually anyone. Across the political spectrum—and with over a dozen political parties, that spectrum runs from hard socialist to neofascist—nearly all party leaders have adopted Denktash's talk of a bicommunal confederation, even if they can't quite articulate what that means. To a degree I've not encountered in any other ethnic conflict zone in the world—not in Bosnia or Sri Lanka, certainly not in Israel—the Turkish Cypriots appear to speak as one, and they have chosen Rauf Denktash to do the talking.

This is not to say, however, that the TRNC stands as some monoracial *Volksland;* rather, it is a place full of quirky little anomalies, reminders of the past that the government has never quite decided whether to tout or be defensive about. In the Karpas Peninsula, the long, thin finger of land that extends to the northeast, some six hundred Greek Cypriots have chosen to remain in their native villages rather than move south, as have a few hundred Maronite Catholics in the western town of Kormakiti; today these stalwarts continue to receive weekly deliveries of "emergency" supplies by United Nations troops. TRNC officials often cite the existence of these communities as proof of their live-and-let-live philosophy but become noticeably fretful at the prospect of a visitor's actually going to them and hearing the residents' litany of complaints against the government.

Throughout the countryside, Greek Orthodox churches have been either boarded up or retrofitted to serve as mosques, and with a frequency that defies coincidence, Orthodox shrines have the bad habit of occupying vitally strategic land, cordoned off behind barbed wire in militarily restricted zones and off limits to all outsiders. With those Greek monuments that the government simply cannot remove from view—like the beautiful little Monastery of Apostolas Varnavas (St. Barnabas) on the Mesaoria plains, one of the most important Orthodox sites on the entire island—they seem to rely on more subtle discouragement; although two major highways in the TRNC pass close by, neither posts signs to the monastery.

To fill up this landscape, with all its vestiges of Hellenistic culture, and to fill up all the formerly Greek villages that were abandoned after the invasion—after all, only 40,000 people moved north to replace the 175,000 who moved south—the TRNC has energetically tried to woo others to move in. Most controversial have been the "Turkish settlers," thousands of peasants from Anatolia, one of the poorest regions in mainland Turkey, who have taken over entire villages on the Mesaoria and built new towns in the flatlands below Famagusta. Socially conservative and largely uneducated, the settlers are looked down upon by the far more liberal and cultured native Turkish Cypriots, and are a source of rage for Greek Cypriots, who see them as interlopers illegally occupying old "Greek land."

At the other extreme are the expatriates, mostly British and Germans, who either have taken up permanent residence in the TRNC or maintain summer homes here, and nowhere is their privileged status more in evidence than in the picturesque village of Karmi. Nestled in the Kyrenia Mountains overlooking Five-Mile Beach, Karmi was a Greek Cypriot village until 1974; today it is "European only" by law, meaning that not just Greeks and mainland Turks are forbidden to own property there but Cypriots as well. Over a game of pool at the cozy Crow's Nest pub, the owner, a good-natured Brit named Steve Clark, explains how that came about.

"Well, once the Turks came ashore in '74, the fuzzies [Greeks] all took off across the mountains—can't say I blame them—and this place just fell apart. A few foreigners were living up here, and they finally got together and went to Denktash and said, 'The only way this village is going to come back is if you make it all European.' Denktash agreed, and that's the way it has been ever since."

Given twenty-five-year leases in return for renovating the village's dilapidated homes, the "Europeans" quickly transformed Karmi into a reasonable facsimile of a Cypriot hill town, if a bit abundant with flower boxes and cute house names. To judge by the minutes of their last town meeting—tacked up in an announcement box on the main square right next to the old Greek church—the residents' most pressing concerns revolve around rising water bills, noisy dogs, and renters who play loud music. Oh, and the ongoing struggle to get their leases extended for another forty-nine years.

"President Denktash has done a lot for us—well, for the whole country," says a slightly hammered Englishwoman at the Crow's Nest, "but we're having a very difficult time getting a clear answer on the leases."

Although many of the other expatriates living along the north coast find the apartheid quality of Karmi distasteful, they share the sentiments of the town's residents in at least one crucial aspect. Like determined expatriates everywhere, there is the tinge of the zealous convert about them. They tend to paint the Cyprus conflict in stark black and white: The Turks have done no wrong, are practically incapable of doing wrong; unification would be "a disaster, a holocaust"; the Greeks are lazy, scheming, vicious, never to be trusted. There is an anger, tinged with racism, to the "Europeans" that one rarely hears among the Turkish Cypriots, and many have directed that anger into lobbying politicians "back home" to grant full recognition to the TRNC, a point that will surely not be lost on President Denktash when the lease extension papers finally reach his desk.

Not surprisingly, the Greek Cypriots have seized on each one of these issues—the desecration of antiquities, the "flood" of Turkish settlers, the "illegal occupations" in Karmi—and added them to their Thousand Points of Plight campaign. For each one, though, Rauf Denktash has a quick and ready response.

As I listen to the president, I begin to wonder how many times he has answered these same questions, given the same lecture—to visiting diplomats, to journalists, to assemblies of his countrymen—and it finally dawns on me why he simply ignored my first question and led me back into history. Because there's really nothing else to talk about. The current situation in Cyprus? Same as last year, same as twenty years ago. Albeit a Greek legend, there is something rather Sisyphean about Rauf Denktash. He has been saying essentially the same thing for twenty-five years, and no one but his choir has listened. The Greek Cypriots, the American and UN peace negotiators who periodically shuttle around the island have always looked for an angle, an opening, and there never has been one. Rauf Denktash is obdurate and unyielding and steeped in history because so are his people.

"Do you ever get tired of this?" I ask. "Hearing the same questions, giving the same answers? Do you ever think of just chucking it all and retiring to Switzerland?"

Denktash slips into a slight smile. "No. I feel it is part of my duty as president to get our message out to the rest of the world in any way I can. Of that I can never tire. And Switzerland is too cold."

At the end of our long interview, as the president is walking me to the door, he suddenly veers over to a high bookcase. Standing on his tiptoes, he reaches up and pulls down an oversize paperback book and hands it to me. It is a collection of the photographs he has taken of his little domain over the years. I quickly leaf through it to show my appreciation—there are some nice portraits of villagers, others that look like standard postcards—and I think of the photograph I've seen of him, his camera strapped around his neck, watching the violent events of August 1996 unfold in the no-man's-land outside Dherinia.

"If the situation in Cyprus was exactly the same fifty years from now," I ask, "would that bother you?"

For the first time, Denktash seems caught slightly off guard. He glances over his bookshelf. "Well, I would like to think that at some point progress would be made, that other nations will recognize our legitimacy."

"But you've found ways to work around that. You have security, you have a homeland. If nothing changed, would it bother you?"

He gives me a shrug. "Not really."

Sebastian Junger
REPUBLIC OF CYPRUS

If you go to Cyprus, pretty soon you will hear about Pyla, a small town outside Larnaca where Turkish and Greek Cypriots live together in peace. The town falls entirely within the buffer zone, so neither side was able to claim it as its own. During the Turkish invasion both sides, at different times, sought protection from the UN, and today they still live together, under the shadow of an UNFICYP observation post. "Together" is a relative term, though. There are two mayors, two town halls, two post offices, two phone systems, and two cafés. There are, in effect, two towns, although Greek Cypriots invariably offer up Pyla as a shining example of bicommunal cooperation.

The other thing Pyla is famous for is fresh fish, a vestige of the black-market trade that once existed in the town. Since the TRNC isn't a recognized country, it may ignore such niceties as import duties and copyright laws, allowing Turkish Cypriot merchants to sell Western knockoffs to Greek Cypriots at rock-bottom prices. Ten years ago Pyla boasted forty or fifty Turkish shops doing a booming business in leather jackets, designer jeans, cheap sunglasses, and basketball shoes, but Greek Cypriot authorities eventually cracked down on the cross-border trade, because any commerce with the TRNC, legitimate or otherwise, was seen as a de facto acceptance of an illegal government and therefore a violation of Greek Cypriot law. Besides, shop owners in Larnaca were losing business. Police started pulling cars over outside Pyla and confiscating illegal goods, and pretty soon the only thing left for sale was fish caught in the TRNC.

I drive to Pyla on a beautiful early-spring day with the tree buds suddenly opening up and the Mediterranean sparkling blue and flawless in the distance. Pyla looks like every other farming town in the area, a cluster of small stone houses and cheap apartment blocks set amid the stubbornly uninteresting fields of eastern Cyprus. There are no checkpoints on the road into town and no policemen to show my papers to, so I just drive in and park in the main square. There is a Greek café on one side, a Turkish café on the other, and a UNFICYP observation tower in the middle. On a nearby hill are a Turkish machine-gun position and a huge metal cutout of Atatürk in profile, striding down the slope into town.

Since there is open access on both sides, Scott has decided to meet

me here for a drink, and as soon as I step out of the car, he comes
walking up and shakes my hand. I'm worried that after a week of
Turkish propaganda he'll start gibbering about Greek atrocities, but he
seems unchanged. He's been here for an hour and has already arranged
an interview with the Turkish mayor, or mukhtar, so we cross the
square and step into a street-level office with a big plate-glass window.
The mukhtar's name is Mehmet Sakali. He wears an old blue suit,
frayed at the cuffs, and a black wool sweater over a shirt and tie. His
shoes need resoling, and he has the kind of leathery skin that you
usually see on farmers or ranch hands. Scott asks how relations are
between the two communities.

"Not so well," he says. "No Turks go to Greek Cypriot coffee
shops and no Greek Cypriots go to Turkish coffee shops. If a Greek
comes and talks to a Turk, the spies in town will interrogate them. Day
by day, they try to keep the people apart."

"How were relations before?"

"They were fine until 1958," the mayor says. "Then EOKA started
killing people."

Scott and I have been told that the UN awarded Pyla a one-
million-dollar renovation grant several years ago, but the town lost the
money because no one could agree on how to spend it. It was an important moment, because a successful collaboration would have served
as a model for the rest of Cyprus. And bicommunal activities, as
they're called, would greatly help the Greek Cypriots' case for being accepted into the European Union, something they have lobbied for energetically over the loud objections of the Turkish Cypriots. Scott asks
him what happened.

"We built a coffee shop and a church with the money," says Sakali,
"but we can't agree on anything else because the Greeks insisted on all
Greek workers. I've worked with three other Greek mukhtars, but
now the Cyprus government is getting into everything and it's no
good. We've set up meetings ten times, and each time this mukhtar has
refused to come or has sent his town clerk. So how can I trust him?"

Scott gives me a baleful stare, which I ignore. After the interview we have a drink at the Greek Cypriot café, and then Scott leaves town and I go to talk to the Greek Cypriot mayor. He's not in, but the town clerk is, a clean-shaven young man named Stavrous Stavron. He offers me a seat in his gleaming new office and asks me what I need to know. I repeat the same questions we asked the Turkish mayor, starting with relations between the two sides.

"It depends on what you're looking at," he says. "You can see neighbors living together peacefully and you can see a village coming into conflict. It's intervention from the outside—by that I mean the politicians—that causes tension. The last year has been very difficult because the new [Turkish] mayor is a protégé of the extremists."

I tell him that the "new mayor"—Sakali—says the deal fell through because the Greek Cypriot mayor kept refusing to meet with him. Stavron shakes his head. "We ended up employing three Greek Cypriots to repair the Orthodox church and twelve Turkish Cypriots to renovate the Turkish coffee shop. Both projects were finished successfully, but then there were elections on the Turkish side and the new mukhtar won without any opposition. The old mukhtar was forced to not be a candidate; that's what I mean by 'outside influence.' "

It seems that Sakali—presumably a puppet of the Denktash regime—sabotaged the project by insisting on complete Turkish control, which of course the Greek Cypriots couldn't accept. After using only one hundred thousand dollars of the million-dollar grant, Pyla had to relinquish the rest because the two sides could not come to an agreement. That each side would pass up nearly a million dollars in order to make the other side look bad is a devastating comment on the political leadership in Cyprus. If they can't cooperate here—in a fully integrated town that is crippled by unemployment—what chance do they have anywhere else?

"The old mukhtar was fair," Stavron adds wistfully. "He was a Turk—we knew he was a Turk; we knew we could never turn him into a Greek—but we appreciated his cooperation."

I thank Stavron for his time and walk back across the square. I have the impression that every person in town knows that Scott and I have been here and that half of them are still watching me through their window slats. I drive out to a Turkish restaurant for some of the fresh fish that Pyla is famous for. The meal is good but not good enough to make a town famous. I eat quickly and get back into the car. Dark clouds are rolling off the Troodos, and by the time I hit the highway a heavy cold rain is washing my windshield.

The Greek Cypriots can never win, I think, racing northwest toward Nicosia. The only thing that will bring stability to the island is a gradual meshing of the economies, and neither side will let that happen. The Greek Cypriots have stubbornly resisted doing any business with the TRNC because that would indirectly support the Denktash regime, and the TRNC has made it an unspoken policy to sabotage any budding relationship between the two countries. If peace came to Cyprus, the Greek Cypriots would become eligible for membership in the European Union, and that is something that Turkey— which has been rebuffed by the EU—could never accept. The only way out for the Greek Cypriots would be to recognize the TRNC diplomatically and declare the hostilities over, but that will never happen. Even acceptance into the EU isn't worth that.

And so the conflict groans on, and the peacekeepers keep walking their patrols.

"All the politicians in the south have been around since before independence," explains a prominent Greek Cypriot journalist (who, to my frustration, months later, requests that he not be identified, a reversal that testifies to the stifling paranoia of Cypriot politics). I seek him out the day after returning from Pyla. "They've made a career out of being defiant," he goes on. "These are the same guys who lost the war in '74. . . . They're prisoners of their own rhetoric; they know fuck-all about anything apart from the Cyprus problem."

The journalist is old enough to be part of the last generation to have any memory of the Turkish invasion. Anyone younger effectively

grew up without contact with Turkish Cypriots and knows only what
the government says about them. Clearly, he is tired of hearing it.

"No one will go on record and say it," he says, "but now your av-
erage man on the street would say, 'Why don't we just build a wall?' "

"Literally build a wall?"

"Yeah, a big wall, them on that side, us on this side," he says.
"And we don't want to see them ever again."

Afterward I walk downtown for lunch. The weather has cleared,
and English tourists are again out in force. They wander in and out of
Gucci and Benetton shops and sit at cafés with their faces turned to
the sun. A few blocks away, thousands of Turkish troops wait in
bunkers for their orders to attack. It'll never happen, I think. They
already have what they want.

Scott Anderson
THE TURKISH REPUBLIC OF NORTHERN CYPRUS

During my last few days in the TRNC, I travel with an interpreter
provided by the government's Office of Public Information. It is an indication of how seriously the government takes its public relations
initiative that the information office falls under the aegis of its
Ministry of Foreign Affairs and Defense, but any concern that Ayshen,
a pleasant, if slightly stiff woman in her mid-thirties, has been assigned
to keep tabs on me is soon dispelled; through most of the interviews,
her boredom is palpable.

As it turns out, Ayshen is originally from the city of Limassol, in
what is now the Republic of Cyprus. Her family was solidly upper-
middle-class—her grandfather a large landowner, her father a
physician—until they lost almost everything in the "enclave era" of

the 1960s. After the 1974 partition they moved north as refugees, and Ayshen eventually went off to attend university in London; she returned to the TRNC only a few years ago, a decision she clearly regrets. "Always it is the same here," she says after one particularly long and tedious day of interviews. "The same politics, the same arguments. Sometimes I feel like I am caught in a nightmare and cannot wake up."

Until last year Ayshen had been active in the "bicommunal" talks initiative. Sponsored by the United Nations and international conflict resolution groups, the talks were designed to bring together small groups of Turkish and Greek Cypriots—businessmen, intellectuals, educators—in hopes that the dialogues might lead to a political opening. Ridiculed by the governments and conservative media of both sides, the effort has largely been abandoned. "It is too bad," Ayshen says, "because I felt it was important that we try anything that might change the situation." Now she carries the label of "peacenik," which causes her some problems around the office.

Between interviews, Ayshen tries to steer me to the more pleasant places to be found in the TRNC, one being the St. Barnabas Monastery outside Famagusta. In the deserted inner courtyard, she sits on a stone bench beneath a bitter lemon tree. "This is one of my favorite places in the whole country," she remarks, "this and the Karpas Peninsula. Up there, it is so quiet—miles of empty beaches, small villages. It's the best place to go to get away from everything."

I know her well enough by now to know what she means by "everything": politics, the speeches, the Problem.

At St. Barnabas Monastery, we are just three or four miles from the Martyred Villages. In late July 1974, a few days after the first phase of the Turkish Peace Operation, EOKA gunmen seized three Turkish villages, led over eighty residents out into the fields to be shot, then threw their bodies in mass graves. Today the road connecting the villages is called Martyr's Way, and beside it are a couple of nearly identical memorial parks, both centered on a stone wall on which the names of the murdered have been carved, both containing freestanding posterboards displaying the same awful photographs of the mass grave exhumations. Along Martyr's Way, large yellow signs, helpfully printed in both English and Turkish, point toward the actual mass graves.

When I mention our proximity to the Martyred Villages, Ayshen's mood falls.

"Do you want to go to them?" she asks.

When I say no, that I've already seen them, she seems tremendously relieved.

On my last day in the TRNC, I convince Ayshen to take me to the village of Tashkent, in the hills just north of Lefkosa. I am curious to see Tashkent, both because of the massive Turkish Cypriot flag that has been painted on the mountainside just outside it and because it is known as the village of widows. The original Tashkent was in the south, and amid the fighting in 1974 nearly all the adult males were rounded up and murdered by Greek Cypriot gunmen. In the population transfer of 1975, the Tashkent widows were brought north and given the formerly Greek village of Vouno as their new home.

Wandering around the village, I spot an old woman in black sitting on her porch on this sunlit day. Her name is Emine Mutallip, and sitting in the sun with her is her ninety-two-year-old father, Mustafa Sadik. Emine graciously brings out chairs for us to sit on and, at my instigation, begins to tell the story of those long-ago killings, the tragedy that took her husband and two brothers. Suddenly both she and her father burst into tears, and then Ayshen does as well.

Afterward, as we wander along Tashkent's main street, Ayshen apologizes for her outburst. "It brings back memories of my own family," she says. "I was just a little girl when I was in the refugee camp, but I remember that we were very poor and I was always hungry. All my family, we had to flee to different places, and always my father was worried, trying to find the others, trying to learn if they were still alive. Three times in my childhood I was a refugee, but I have not thought about it for a long time."

As we drive down the long hill toward Lefkosa, however, I discover that there's another reason for Ayshen's unhappiness. Some ten days earlier, on the very day I arrived in the TRNC, Turkish security agents in Nairobi grabbed Abdullah Ocalan, the leader of the militant Kurdistan Worker's Party, and whisked him back to Turkey. Ocalan, a man the Turkish government holds responsible for the deaths of some thirty thousand Turks and whom the United States government has classified as a terrorist, had been harbored in Kenya by the Greek embassy and had been traveling on a doctored Republic of Cyprus passport.

"For me, I think that is the end," Ayshen says in the car. "Before, I think I didn't want to believe how much the Greeks hated us, that maybe there was a way for us to live together again. But for them to support a man like Ocalan just because he kills Turks, now I see how much they hate us. Now I cannot see any way out of this."

As I drive, I think of what a hopeless, bitter place this is. Cyprus is like some boat sunk under a great weight of stones, and while the rest of the world talks of finding some way to refloat it, none of the stones is ever removed. Instead, the Greeks and Turks busy themselves finding more stones to drop onto the wreck: the Dherinia killings, the struggle over European Union membership, the Ocalan affair. Tomorrow, no doubt, they'll find another.

So how do you fix it? Both sides in this conflict wield history as a weapon and invoke it as the basis for their own plaintive cry for justice. But if the history of Cyprus—indeed, the history of most of the world—reveals anything, it is that there is no such thing as justice: You live in your house until the day someone comes along and throws you out, and then he lives there until someone else comes along to throw him out. Just where do you pinpoint the moment in this island's history and say, "Here, we will right this wrong," and let all the previous ones go by the wayside? Obviously, you cannot afford to go very far back, because in Cyprus, as everywhere else, there is always a prior victim.

More specifically, how do you fix it when both sides clearly have so little interest in doing so themselves? Start small, I suppose. Point out to them that wallpapering their countryside with grisly photos of those killed by the other side may not be the best way to foster fraternal thoughts. Suggest that it might be imprecise to describe a military offensive in which thousands were killed as a "peace operation," or that there may be a better way to bring one's rivals to the negotiation table than by referring to them as "the so-called ministers of the pseudo-state." Even these baby steps the Cypriots will not take. By steadfastly clinging to the rhetoric of a quarter century ago, by stoutly refusing to make any concession, you finally have to conclude that it's because they want it this way.

But there is, perhaps, another way to look at all this. In the fifteen years of ethnic violence before the 1974 partition, hundreds of Cypriots on both sides were killed. In the twenty-five years since, there have been a total of sixteen—or about the same number that die on the island's highways in a bad month. At a cost of ninety million dollars a year, the United Nations has brought calm to an intractable conflict zone—about what the recent NATO military operations in Kosovo cost for just two days. Of course, people have suffered and lost a great deal in Cyprus—especially all those uprooted from their homes and forced to start over again—but at least now, kept apart by the buffer, they have been given the chance to start over. That's far better than what usually happens in war zones.

So perhaps what has passed as "The Cyprus Problem" all these years has actually been "The Cyprus Solution," and perhaps the diplomats who periodically wring their hands over the ongoing stalemate on this island should actually be taking notes and trying to export it elsewhere. That would require new thinking among the power brokers of the West, and perhaps especially among those in Washington, embroiled in the latest crisis in the Balkans. Maybe what most needs to end is all the chatter about exit strategies. Those in power must recognize that there is no exit from bad history and that at certain times and in certain places the best that can be done is to simply stand between the fighters indefinitely and hope that someday they'll get over it and move on—not in a year, not in ten years, but maybe eventually. Until then the least costly solution, in terms of both blood and money, is to give the Bosnians and Serbs and Kosovars of the planet what the Cypriots already have, a "dead zone" across which they can hurl accusations and threats in safety. At least it will give them something to talk about, and all that the rest of the world will have to suffer is the hearing of it.

Back in Lefkosa I leave Ayshen at the entrance to the Office of Public Information and watch her walk slowly, head bowed in sadness, up the entranceway. It occurs to me that it is the people like her—the earnest, the "peaceniks," the goodhearted and forgiving—who are the last, quiet victims of this place. They are to be found in Bosnia and Serbia and Kosovo as well, of course, those who refuse to believe that a culture once torn apart can't be put back together again, who forever wait for their day to come.

COLTER'S WAY

1999

Late in the summer of 1808 two fur trappers named John Colter and John Potts decided to paddle up the Missouri River, deep into Blackfeet territory, to look for beaver. Colter had been there twice before; still, they couldn't have picked a more dangerous place. The area, now known as Montana, was blank wilderness, and the Blackfeet had been implacably hostile to white men ever since their first contact with Lewis and Clark several years earlier. Colter and Potts were working for a fur trader named Manuel Lisa, who had built a fort at the confluence of the Yellowstone and Bighorn rivers. One morning in mid-August they loaded up their canoes, shoved off into the Yellowstone, and started paddling north.

Colter was the better known of the two men. Tall, lean, and a wicked shot, he had spent more time in the wilderness than probably any white man alive—first as a hunter on the Lewis and Clark Expedition, then two more years guiding and trapping along the Yellowstone. The previous winter he'd set out alone, with nothing but a rifle, a buffalo-skin blanket, and a thirty-pound pack, to complete a several-month trek through what is now Montana, Idaho, and Wyoming. He saw steam geysers in an area near present-day Cody, Wyoming, that was later dubbed Colter's Hell by disbelievers. Within weeks of arriving back at Lisa's fort in the spring of 1808, he headed right back out again, this time up to the Three Forks area of Montana, where he'd been with Lewis and Clark almost three years earlier. His trip was cut short when he was shot in the leg during a fight with some Blackfeet, and he returned to Lisa's fort to let the wound heal. No sooner was he better, though, than he went straight back to Three Forks, this time with John Potts. The two men quickly amassed almost a ton of pelts, but every day they spent in Blackfeet territory was pushing their luck. Finally, sometime in the fall, their luck ran out.

As they paddled the Jefferson River, five hundred Blackfeet Indians suddenly swarmed toward them along the bank. Potts grabbed his rifle and killed one of them with a single shot, but he may have done that just to spare himself a slow death; the Blackfeet immediately shot him so full of arrows that "he was made a riddle of," as Colter put it. Colter surrendered and was stripped naked. One of the Blackfeet asked whether he was a good runner. Colter had the presence of mind to say no, so the Blackfeet told him he could run for his life; when they caught him, they would kill him. Naked, unarmed, and given a head start of only a couple of hundred yards, Colter started to run.

He was, as it turned out, a good runner—very good. He headed for the Madison River, six miles away, and by the time he was halfway there, he'd already outdistanced every Blackfoot except one. His pursuer was carrying a spear, and Colter spun around unexpectedly, wrestled it away from him, and killed him with it. He kept running until he got to the river, dived in, and hid inside a logjam until the Blackfeet got tired of looking for him. He emerged after nightfall, swam several miles downstream, then clambered out and started walking. Lisa's fort was nearly two hundred miles away. He arrived a week and a half later, his feet in shreds.

Clearly, Colter was a man who sought risk. After two brutal years with Lewis and Clark, all it took was a chance encounter with a couple of itinerant trappers for Colter to turn around and head back into Indian territory. And the following summer—after three straight years in the wild—Manuel Lisa convinced him to do the same thing. Even Colter's narrow escape didn't scare him off; soon after recovering from his ordeal, he returned to the Three Forks area to retrieve his traps and had to flee from the Blackfeet once again. And in April 1810 he survived another Blackfeet attack on a new stockade at Three Forks, an attack that left five men dead. Finally Colter had had enough. He traveled down the Missouri and reached St. Louis by the end of May. He married a young woman and settled on a farm near Dundee, Missouri. Where the Blackfeet had failed, civilization succeeded: He died just two years later.

Given the trajectory of Colter's life, one could say that the wilderness was good for him, kept him alive. It was there that he functioned at the outer limits of his abilities, a state that humans have always thrived on. "Dangers . . . seemed to have for him a kind of fascination," another fur trapper who knew Colter said. It must have been while under the effect of that fascination that Colter felt most alive, most potent. That was why he stayed in the wilderness for six straight years; that was why he kept sneaking up to Three Forks to test his skills against the Blackfeet.

Fifty years later, whalers in New Bedford, Massachusetts, would find themselves unable to face life back home and—as miserable as they were—would sign up for another three years at sea. A hundred years after that, American soldiers at the end of their tours in Vietnam would realize they could not go back to civilian life and would volunteer for one more stint in hell.

"Their shirts and breeches of buckskin or elkskin had many patches sewed on with sinews, were worn thin between patches, were black from many campfires, and greasy from many meals," writes historian Bernard De Voto about the early trappers. "They were threadbare and filthy, they smelled bad, and any Mandan had lighter skin. They gulped rather than ate the tripes of buffalo. They had forgotten the use of chairs. Words and phrases, mostly obscene, of Nez Percé, Clatsop, Mandan, Chinook came naturally to their tongues."

None of these men had become trappers against his will; to one degree or another, they'd all volunteered for the job. However rough it was, it must have looked better than the alternative, which was—in one form or another—an uneventful life passed in society's embrace. For people like Colter, the one thing more terrifying than having something bad happen must have been to have nothing happen at all.

Modern society, of course, has perfected the art of having nothing happen at all. There is nothing particularly wrong with this except that for vast numbers of Americans, as life has become staggeringly easy, it has also become vaguely unfulfilling. Life in modern society is designed to eliminate as many unforeseen events as possible, and as inviting as that seems, it leaves us hopelessly underutilized. And that is where the idea of "adventure" comes in. The word comes from the Latin *adventura,* meaning "what must happen." An adventure is a situation where the outcome is not entirely within your control. It's up to fate, in other words. It should be pointed out that people whose lives are inherently dangerous, like coal miners or steelworkers, rarely seek "adventure." Like most things, danger ceases to be interesting as soon as you have no choice in the matter. For the rest of us, threats to our safety and comfort have been so completely wiped out that we have to go out of our way to create them.

About ten years ago a young rock climber named Dan Osman started free-soloing—climbing without a safety rope—on cliffs that had stymied some of the best climbers in the country. Falling was not an option. At about the same time, though, he began falling on purpose, jumping off cliffs tethered not by a bungee cord but by regular climbing rope. He found that if he calculated the arc of his fall just right, he could jump hundreds of feet and survive. Osman's father, a policeman, told a journalist named Andrew Todhunter, "Doing the work that I do, I have faced death many, many, many times. When it's over, you celebrate the fact that you're alive, you celebrate the fact that you have a family, you celebrate the fact that you can breathe.

Everything, for a few instants, seems sweeter, brighter, louder. And I think this young man has reached a point where his awareness of life and living is far beyond what I could ever achieve."

Todhunter wrote a book about Osman called *Fall of the Phantom Lord*. A few months after the book came out, Osman died on a twelve-hundred-foot fall in Yosemite National Park. He had rigged up a rope that would allow him to jump off Leaning Tower, but after more than a dozen successful jumps by Osman and others, the rope snapped and Osman plummeted to the ground.

Colter of course would have thought Osman was crazy—risk your life for no good reason at all?—but he certainly would have understood the allure. Every time Colter went up to Three Forks, he was in effect free-soloing. Whether he survived or not was entirely up to him. No one was going to save him; no one was going to come to his aid. It's the oldest game in the world—and perhaps the most compelling.

The one drawback to modern adventuring, however, is that people can mistake it for something it's not. The fact that someone can free-solo a sheer rock face or balloon halfway around the world is immensely impressive, but it's not strictly necessary. And because it's not necessary, it's not heroic. Society would continue to function quite well if no one ever climbed another mountain, but it would come grinding to a halt if roughnecks stopped working on oil rigs. Oddly, though, it's the mountaineers who are heaped with glory, not the roughnecks, who have a hard time even getting a date in an oil town. A roughneck who gets crushed tripping pipe or a fire fighter who dies in a burning building has, in some ways, died a heroic death. But Dan Osman did not; he died because he voluntarily gambled with his life and lost. That makes him brave—unspeakably brave—but nothing more. Was his life worth the last jump? Undoubtedly not. Was his life worth living without those jumps? Apparently not. The task of every person alive is to pick a course between those two extremes.

I have only once been in a situation where everything depended on me—my own version of Colter's run. It's a ludicrous comparison except that for the age that I was, the stakes seemed every bit as high. When I was eleven, I went skiing for a week with a group of boys my age, and late one afternoon when we had nothing to do, we walked off into the pine forests around the resort. The snow was very deep, up to our waists in places, and we wallowed through slowly, taking turns breaking trail. After about half an hour, and deep into the woods now, we crested a hill and saw a small road down below us. We waited a few minutes, and sure enough, a car went by. We all threw snowballs at it. A few minutes later another one went by, and we let loose another volley.

Our snowballs weren't hitting their mark, so we worked our way down closer to the road and put together some really dense, heavy iceballs—ones that would throw like a baseball and hit just as hard. Then we waited, the woods getting darker and darker, and finally in the distance we heard the heavy whine of an eighteen-wheeler downshifting on a hill. A minute later it barreled around the turn, and at the last moment we all heaved our iceballs. Five or six big white splats blossomed on the windshield. That was followed by the ghastly yelp of an air brake.

It was a dangerous thing to do, of course: The driver was taking an icy road very fast, and the explosion of snow against his windshield must have made him jump right out of his skin. We didn't think of that, though; we just watched in puzzlement as the truck bucked to a stop. And then the driver's side door flew open and a man jumped out. And everyone started to run.

I don't know why he picked me, but he did. My friends scattered into the forest, no one saying a word, and when I looked back, the man was after me. He was so angry that strange grunts were coming out of him. I had never seen an adult that enraged. I ran harder and harder, but to my amazement, he just kept coming. We were all alone in the forest now, way out of earshot of my friends; it was just a race between him and me. I knew I couldn't afford to lose it; the man was too crazy, too determined, and there was no one around to intervene. I was on my own. *Adventura*—what must happen will happen.

Before I knew it, the man had drawn to within a few steps of me. Neither of us said a word; we just wallowed on through the snow, each engaged in our private agonies. It was a slow-motion race with unimaginable consequences. We struggled on for what seemed like miles but in reality was probably only a few hundred yards; the deep snow made it seem farther. In the end I outlasted him. He was a strong man, but he spent his days behind the wheel of a truck— smoking, no doubt—and he was no match for a terrified kid. With a groan of disgust he finally stopped and doubled over, swearing at me between breaths.

I kept running. I ran until his shouts had died out behind me and I couldn't stand up anymore, and then I collapsed in the snow. It was completely dark and the only sounds were the heaving of the wind through the trees and the liquid slamming of my heart. I lay there until I was calm, and then I got up and slowly made my way back to the resort. It felt as if I'd been someplace very far away and had come back to a world of tremendous frivolity and innocence. It was all lit up, peals of laughter coming from the bar, adults hobbling back and forth in ski boots and brightly colored parkas. "I've just come back from some other place," I thought. "I've just come back from some other place these people don't even know exists."

THE FORENSICS OF WAR

1999

Homo homini lupus. (Man is a wolf to man.)
—PLAUTUS, *Asinaria*

No one knows who he was, but he almost got away. He broke and ran when the Serbs started shooting, and he made it to a thicket before the first bullet hit him in the left leg. It must have missed the bone, because he was able to keep going—along the edge of a hayfield and then into another swath of scrub oak and locust. There was a dry streambed in there, and he probably crouched in the shadows, listening to the bursts of machine-gun fire and trying to figure out a way to escape. The thicket stretched uphill, along the hayfield, to a stand of pine trees, and from there it was all woods and fields leading to the Albanian border. It didn't offer much of a chance, and he must have known that.

He tied a sweater around the wound in his thigh and waited. Maybe he was too badly hurt to keep moving, or maybe he didn't dare because the Serbs were already along the edge of the field. Either way, they eventually spotted him and shot him in the chest, and he fell backward into the streambed. His killers took his shoes, and—months later, after the war ended—a fellow Albanian took his belt buckle and brought it to the authorities in Gjakovë. It was the only distinctive thing on him, and there was a chance that someone might recognize it.

I saw the dead man in late June, two weeks after NATO had taken Kosovo from the Serbs. It was a hot day, and my photographer and I stood peering at his corpse, in the same mottled shade that the man had tried to hide in. His skull was broken open, and his jawbone was a short distance away. The sweater was still tied around his leg. I had walked into the thicket braced for the worst, but he wasn't particularly hard to look at. He'd been killed two months earlier—on April 27, around midday—and he looked less like a person than a tipped-over hatrack draped in blue jeans and a cheap parka. The young man who had led us there leaned on a shepherd's crook and told us that the dead man was in his early twenties and had probably come from a nearby village. The Serbs had swept the valley from Junik to Gjakovë in retaliation for an attack by the Kosovo Liberation Army, which for two years has fought for independence for Kosovo. They'd taken the men from more than half a dozen little villages and gunned them down in a field outside Meja. Then they came back in the middle of the night to bury them. They missed a few.

The shepherd identified himself as Bashkim; he was a handsome blond kid with a wispy goatee and a shy smile that never left his face. "They came at five A.M.—not shooting, just yelling," Bashkim said. "They made two hundred people lie down against a compost heap, piled cornhusks on them, and then machine-gunned them. Then they set them on fire. . . . It was local militia from Gjakovë. They were wearing green camouflage and black ski masks. One of them was called Stari; all the women saw him. They recognized him from Gypsy Road, about five kilometers from here."

Meja was just a scattering of tile-roofed farmhouses along a dirt road in the middle of a broad agricultural valley. Wheat and hayfields gave way to brush-covered hills and then the Koritnik Mountains, which run along the Albanian border. Bashkim had escaped the roundup of men in the valley because he was in an isolated house that the Serbs missed. While telling the story, he seemed undisturbed by the massacres, his own close call, or even the body at his feet. He just kept smiling and smoking the American cigarettes we offered him. After twenty minutes or so, he led us back into the hot sun of the hayfield and past the compost heap where the men had been shot. There was a human leg in the grass, and then another leg, and then more remains in a ditch. They were harder to identify. Stuck into the compost heap was an old umbrella. "Why is that there?" I asked.

"It was found near one of the bodies," Bashkim said. "Maybe someone will recognize it and know who he was."

The worst of the violence didn't come to southwestern Kosovo until the evening of March 24, when NATO jets streaked overhead on their way to bomb command and control targets in Serbia. Within hours Serb special police, soldiers, and hastily deputized militia units were walking through the streets of nearby Gjakovë, pumping incendiary rounds, known locally as butterflies, into houses and storefronts in the Albanian part of town. When the buildings finished burning, the Serbs knocked the walls over with bulldozers and then used Gypsies to clear the rubble from the streets so that they were passable for tanks. Anyone who stood around and watched was shot.

There was little the KLA could do but hide in the hills and wait for it to be over. In two years of fighting, the KLA never won a battle or held a town for long, but it did know how to ambush. And in mid-April, just outside Meja, it pulled off an ambush that would bring the full wrath of the Serbs down on the valley.

Its target was a Serb commander named Milotin Prasović, who was particularly loathed by the local Albanians. A week or so earlier, Prasović had driven through Meja warning the residents that he was going to return to collect all the weapons in town, and if there weren't any for him, he'd burn their houses down. True to his word, he came back in a brick-red Audi filled with police. They drove through town, shot into the air, turned around, and drove straight into a KLA ambush. The first rocket-propelled grenade blew the right rear door off. That was followed by another round and sheets of automatic-weapons fire. Everyone in the car was killed except for Prasović, who managed to dive out of his seat and start shooting back from the edge of the road. It was over within seconds; the KLA shot him down from their hiding place and then retreated into the hills above town.

The retaliation, when it came, was swift and implacable. Shortly before dawn on April 27, according to locals, a large contingent of Yugoslav troops garrisoned in Junik started moving eastward through the valley, dragging men from their houses and pushing them into trucks. "Go to Albania!" they screamed at the women before driving on to the next town with their prisoners. By the time they got to Meja they had collected as many as three hundred men. The regular army took up positions around the town while the militia and para-militaries went through the houses grabbing the last few villagers and shoving them out into the road. The men were surrounded by fields most of them had worked in their whole lives, and they could look up and see mountains they'd admired since they were children. Around noon the first group was led to the compost heap, gunned down, and burned under piles of cornhusks. A few minutes later a group of about seventy were forced to lie down in three neat rows and were machine-gunned in the back. The rest—about thirty-five men—were taken to a farmhouse along the Gjakovë road, pushed into one of the rooms, and then shot through the windows at point-blank range. The militiamen who did this then stepped inside, finished them off with shots to the head, and burned the house down. They walked away singing.

———

By conservative estimates, the Serbs killed at least ten thousand people in Kosovo. There are so many bodies—both human and animal—lying around the countryside that much of the rural water supply is contaminated. There are parts of Kosovo where not one village has been spared, and there are villages where not one house has been left standing. In the Decani district, bodies have been dumped in the wells of thirty-nine of forty-four villages surveyed. When NATO tanks rumbled into Kosovo on June 12, they found a level of destruction that hadn't been seen in Europe since World War II.

The first big massacre occurred in March 1998, when Serb forces surrounded the village of Prekaz and wiped out fifty-eight civilians, many of them women and children. The attack was in retaliation for a shoot-out between KLA and Serb police a couple of weeks earlier, and it was the beginning of a horrible symbiosis between the two forces. Every time the KLA carried out a guerrilla attack, Serb forces would destroy the nearest village and massacre as many of the inhabitants as they could. And every time the Serbs massacred people in a village, more grief-stricken survivors joined the KLA. "For every massacre Serbs commit, we get twenty more recruits," one KLA commander told a journalist friend of mine a few weeks before the NATO bombings started.

Ethically there's an extremely thin line between ambushing Serb forces and deliberately provoking Serbs into massacring civilians, but the strategy worked. Around 8:00 P.M. on March 24, the first NATO warplanes struck targets deep within Serbia. And two months later, on May 24—just as the first sketchy peace agreements were being explored with Belgrade—the Hague war crimes tribunal indicted Slobodan Milošević and 4 other government and military leaders. The indictment was based on eyewitness accounts of massacres that took place between January and April 1999 in the villages of Racak, Krushe e Mahde, Krushe e Vogel, Bellacerka, Izbica, and Padalishte; the indictment listed, by name, more than 340 ethnic Albanians who had been killed. Within days of NATO's arrival in Priština, war crimes investigators donated by many NATO countries were sifting through the mass graves named in the indictments, gathering evidence.

The crimes that Milošević and his compatriots were charged with fall under the Geneva Conventions of 1949, which were a direct outgrowth of the post–World War II Nuremberg trials. When the Germans surrendered on May 7, 1945, the Allies were suddenly faced with an unprecedented problem: They had in their custody Nazi officials who had started a war in which nearly fifty million people had been killed. Many of the dead had been exterminated in concentration camps, and the question was: What kind of justice should be brought to bear on the men who carried out such slaughter? The British initially suggested that the hundred or so main German culprits simply be taken out into the woods and shot (an idea embraced by Joseph Stalin, who jokingly—or maybe not—proposed upping the number to fifty thousand). Ultimately, though, due process prevailed. The accused would be given trials, which "they, in the days of their pomp and power, never gave to any man," as the chief American prosecutor, Robert Jackson, put it. The trial would be open and fair, conducted in both English and German, and the accused would be represented by lawyers who would call their own witnesses and cross-examine others.

As idealistic as it was, the idea had inherent flaws. First, it was, by definition, a victor's justice, and there was no suggestion that the victors would ever face the same scrutiny as the vanquished. The Soviets, for example, had invaded and occupied eastern Poland in 1939, in close cooperation with the Nazis; they massacred thousands of Polish officers and buried their bodies in the Katyn Forest. The Allies had fire-bombed Dresden, killing several hundred thousand civilians, and the Americans had fire-bombed Tokyo and then dropped nuclear bombs on Hiroshima and Nagasaki. These were all direct attacks on civilians—and therefore clear violations of international law—but they would never make it to the docket at Nuremberg.

Second, the Nazis were charged with, among other things, crimes against humanity, which include crimes committed by a government against its own people. At the start of World War II the law didn't exist, and because the Holocaust was completely legal under German law, the perpetrators had technically never committed a crime. To charge them ex post facto was illegal and would never have stood up in a regular judicial proceeding.

These objections amounted to legal parlor games, however; the reality was that the Nuremberg trials were about as fair as things ever get in wartime. Out of the twenty-two Nazi leaders who were tried, twelve were sentenced to hang—including Reichsmarschall Hermann Göring, who swallowed a cyanide pill shortly before his execution; seven were sentenced to long prison terms; and three were acquitted. Three years later, the legal principles used in the trials were codified as the four Geneva Conventions and the Genocide Convention. Along with the Additional Protocols of 1977, they form the basis today of international war crimes trials. Because they are rooted in something called customary international law, which flows from norms evolved over centuries, rather than from treaties, the conventions are binding even on nations that have not signed them. A state, in other words, cannot exclude itself from the constraints of customary law any more than an individual can.

The indictments announced in The Hague on May 27 charged Yugoslav President Slobodan Milošević, Serb President Milan Milutinović, Deputy Prime Minister Nikola Sainović, Chief of Staff Dragoljub Ojdanić, and Serb Minister of Internal Affairs Vlajko Stojilković with three counts each of crimes against humanity and one count each of violation of the laws and customs of war. Copies of the arrest warrants were sent to all member states of the United Nations and the Yugoslav minister of justice; UN member states were asked to freeze the assets of the accused. The announcement of the indictments was delayed until representatives from international agencies could safely leave the former Yugoslavia, and eyewitnesses wouldn't be identified until the accused were arrested. They could then be properly sheltered from threat and intimidation.

Of the two charges, violation of the laws and customs of war is the older and more traditional. It attempts to reconcile human suffering with the need of an army to defeat its enemy. Although constraints on wartime behavior date back to ancient Hindu and Greek law, the first European wasn't tried in a civilian court until the late fifteenth century, when an Austrian nobleman named Peter von Hagenbach was sentenced to death for atrocities committed under his command. A hundred and fifty years later a Dutch lawyer named Hugo Grotius wrote *The Law of War and Peace,* which is considered the foundation of modern humanitarian law. "Throughout the Christian world . . . I observed a lack of restraint in relation to war, such as even barbarous races would be ashamed of," Grotius wrote. "[A] remedy must be found . . . that men may not believe either that nothing is allowable, or that everything is."

Modern laws and customs of war are direct descendants of Grotius's work. In essence, these laws acknowledge that death and suffering are inevitable in armed conflict, but that deliberately inflicting unnecessary suffering is a criminal act for which individuals can be held accountable. If you shell a military base and happen to kill civilians, you have not committed a war crime; if you deliberately target cities and towns, you have. Killing prisoners, civilians, or hostages is a war crime, as are enslavement of civilians, deportation, plunder, wanton destruction, and "extensive destruction not justified by military necessity."

What is and isn't justified by military necessity is, naturally, open to interpretation. One of the key concepts, though, is the law of proportionality. A military attack that results in civilian casualties—"collateral damage"—is acceptable as long as the military benefits outweigh the price that is paid by humanity. A similar concept is applied to weapons. No matter how many people you kill, using a machine gun in battle is not a war crime because it does not cause unnecessary suffering; it simply performs its job horrifyingly well. Exploding bullets, on the other hand, which were banned in the St. Petersburg Declaration of 1868, mutilate and maim foot soldiers without conferring any additional advantage to the other side. A wounded soldier is usually put out of commission when he is hit. There is no reason to maximize his suffering by using an exploding bullet.

Despite the graceful logic of these principles, warfare remains a chaotic business that will always resist governments' efforts to legislate it. Still, it is all too clear that Serb forces in Kosovo violated just about every law in the book. Furthermore, the violations were carried out on such a massive scale that they also qualified as crimes against humanity—that is, they represented a widespread and systematic campaign against a particular population. Massacring noncombatants—as countless armies have done, including our own—is simply a war crime; trying to drive an entire group of people from your country is a crime against humanity. One could reasonably argue that the Turkish pogrom against the Armenians during World War I qualifies as a crime against humanity, as does the United States' ethnic cleansing of Native Americans.

The novel thing about the 1949 humanitarian law was that it protected the citizens of an offending state as well as those of foreign states, and it applied to peacetime as well as war. Until then governments could do pretty much what they wanted with their own citizens; the most that another nation could do was express its "concern" over the situation. After 1949, in theory a government's conduct at home would be subject to the same standards as its conduct abroad, and human rights abuses could no longer be dismissed as simply an "internal matter." National sovereignty, in other words, would never again protect a government from the bite of the law.

The Hague's press spokesman in Kosovo after NATO's takeover was a young Englishman named Jim Landale. Dressed in jeans and a fleece jacket, with a backpack over his shoulder, he looked more than anything like a college student striding around the streets of Priština. After returning from Meja, I found him in the new UN headquarters, a concrete office building behind the huge, ghastly Grand Hotel at the center of town. A nearby apartment block was smoking slowly from an arson fire, and a lot of young ethnic Albanians were hanging out in front of the hotel, looking for work with the foreign journalists. Landale and I crossed the street to one of the cafés that had just opened and ordered the last two beers that it had in stock.

The tribunal's full name is the International Criminal Tribunal for the Former Yugoslavia—or ICTY, as it's commonly known. It was created in May 1993 to prosecute Bosnian war crimes, and it was later joined by another tribunal for Rwanda. The ICTY has successfully prosecuted several cases from the Bosnian War, mostly against Serbs, despite being hampered by evidence that was in some cases several years old. But it has never before investigated war crimes that were committed so recently, and it has never indicted a head of state during an ongoing armed conflict. Landale, between interruptions by cell phone, explained to me the ICTY strategy in Kosovo.

"This is our biggest undertaking by far," he said. "We're still trying to assess and prioritize all the sites. The murder of civilians is a war crime . . . what we'd try to assess would be: How do the various factors relate to our investigations for indictments against certain individuals? . . . Most villages have some sort of crime scene; every one's got their massacre site just around the corner. To be realistic, we are not going to be able to visit every site in Kosovo."

The criteria for prioritizing sites are unflinchingly pragmatic. One would think that Meja, where several hundred men were machine-gunned in a field, would make a better investigation site than, say, a house where just one family was wiped out. Or that the murder of twenty women and children in a basement would be easier to prosecute than the summary execution of twenty KLA soldiers after a battle. Not so; for the most part, sites are chosen simply for the quality of corroborating evidence that can be gathered. During the NATO bombing campaign, investigators in the refugee camps systematically recorded hundreds of eyewitness accounts of massacres; it was information from those interviews that led to the original indictment handed down in May. A small site with excellent eyewitness accounts, in the eyes of a war crimes prosecutor, is far more valuable than a large site with none.

Similarly, a massacre in an area where a certain military or paramilitary group—such as Arkan's Tigers, who allegedly committed numerous massacres around Gjakovë—was known to have been working is higher priority than a massacre site where the killers are unknown. And a site that presents any sort of access problem is quickly superseded by one that can be worked on immediately. Land mines are considered an access problem, as are remote areas and sites with too many bodies. The amount of labor required to dig up even one body is considerable, so a mass grave like the one in Meja requires backhoes and bulldozers. The first crime teams flown into Kosovo didn't think to bring any.

"One of our hopes is that the tribunal will work as a deterrent," said Landale when I asked him—given the chance that Milošević will never be brought to trial—what the point of all this is. "And it will also help to relieve a collective sense of guilt. Not all people on one side are guilty—just certain individuals. The investigations should make that clear."

There are more than a dozen crime teams in Kosovo, totaling some three hundred people. Scotland Yard, the Royal Canadian Mounted Police, and the FBI all have sent teams, as have police agencies from Germany, Denmark, France, Holland, and Switzerland. Their job is to photograph and diagram massacre sites named in the

ICTY indictments and to gather such evidence as shell casings, bullets, bloody clothing, and anything else that might identify the killers and the method of death. Then the teams attempt to identify human remains—at least to the extent of knowing the age and sex—and conduct autopsies to determine the cause of death. The large mass graves with hundreds of bodies will be investigated months later, after the delicate surface evidence all has been gathered. Buried corpses change very slowly. Most of the evidence they contain is still there a year later.

The FBI team, consisting of sixty-four people and 107,000 pounds of equipment, arrived in Skopje on June 22 in an air force C-5. The team was completely self-sufficient, and it included, on loan from the Armed Forces Institute of Pathology, a forensic anthropologist, two forensic pathologists, and a criminalist, as well as FBI evidence collection experts, two caseworkers from Physicians for Human Rights, a trauma surgeon, and heavily armed Hostage Response Team agents. From Skopje, the team continued by marine helicopter and truck convoy to Gjakovë, where they set up their tents and field morgue under some shade trees inside the Italian Army field base. On the flight over, they got their first glimpse of Kosovo's devastation: entire villages burned to the ground and cows lying in fields, their hindquarters blown off by land mines. Early the next morning in Gjakovë, they got to work.

They had to go no farther than across the street to investigate one of their first crime scenes, a house where Serb special police had executed six men in the middle of the night. A seventh man was wounded but didn't die. He managed to crawl out of the house as it burned and—with the help of female relatives—eventually make it to Albania before dying from loss of blood.

The FBI quickly moved through three more sites in Gjakovë, including one where Serbs had taken twenty-five men at gunpoint and mowed them down with machine-gun fire. There were no survivors, but there were eyewitnesses, and again, their accounts were recorded by ICTY investigators. The ICTY was running into a problem, though: There were so many bodies in Kosovo that every time it investigated a site, locals would tell the investigators about several more, and the list was growing almost exponentially. The FBI team, which had originally been charged with investigating only two sites, worked so quickly that the ICTY tacked seven more on. Most were in and around Gjakovë, and two were outside Peja, in the northwestern corner of the province.

The more remote of the two sites near Peja was known as the Well, outside a little village named Studenica. Around midday on April 12, Serb paramilitaries identifying themselves as Arkan's Tigers executed nine people at a farmhouse and then dumped the bodies down the well. They then smashed the stone and mortar wall surrounding it and dropped the rubble down the hole. Two months later villagers returning to the area dug the well out and pulled up nine badly decomposed bodies. Eight of them were buried at some distance, but one—that of eighty-six-year-old Sali Zeqiraj—was buried in the yard in front of the farm.

I drove up to Studenica early in the morning with the FBI convoy. It was a beautiful spot, smack up against the Koritnik Mountains, with all Kosovo spread out below us. Plumes of smoke rose over the valley from Serb houses that had been set on fire by the KLA. The field had not been de-mined, so the FBI investigators climbed carefully out of their army vehicles and approached the farmhouse along the tire ruts that had been left by previous cars. Roger Nisley, leader of the mission, went ahead to scout the house out and then gathered the investigators in front of the lead Humvee. "It looks like shots were fired through the window," he said. "So get a sketch of that. And we have a body here, apparently it's the grandfather, but we have permission to dig up only one body. There is a total of nine."

While evidence collection people photographed the house and picked through it for shell casings and bullet holes, four diggers started opening up the grave. Family members stood in the field, anxiously stripping grass stems—just far enough away that they didn't have to see anything they didn't want to see. Only one of them—a young man named Xhevat Gashi, who lived in Germany and had come back only the day before—stood close by. The body came up swaddled in clear plastic sheeting, tied with rope at both ends. The investigators, dressed in white Tyvex jump suits with face masks that pinched their nostrils closed, untied the rope and carefully unwrapped the body. The dead man was dressed in pants and socks and a plaid shirt; the investigators cut the clothes off him and laid him out on a blue tarp. He'd been in water for two months, and his flesh had a doughy look, as if he were a mannequin made out of bread. It was very hard to make a connection between the body on the tarp and a human being. Out in the field, one of the family members started to cry. Xhevat, the grandson, shifted on his feet and kept watching.

The investigators quickly found something of interest. "We have an entrance gunshot wound in the back of the head," said Dr. Andrew Baker, one of the pathologists on loan from the Armed Forces Institute of Pathology. "You've got a sharp edge on the outside, blunt edge on the inside. I don't see an exit wound so far."

The fact that there was no exit wound was important because it meant that the bullet—a crucial piece of evidence—might still be in the skull. Baker made an incision in the back of the scalp and pulled the skin forward until he had peeled the face down like a thick rubber mask. Then he opened the skull and probed inside the cavity. It took fifteen minutes to search the skull and the countless folds of the brain; Baker found neither the bullet nor an exit wound. He pulled the face back up, reassembled the head, and rewrapped the dead man in plastic. Then the four diggers lowered him back into the grave.

One of the advantages to investigating a shooting murder (as opposed to a knifing or a bludgeoning) is that ballistics is a precise science and bullets act in fairly predictable ways. By reconstructing the path of a bullet—through a room or through a body—it is possible to know a lot about where it was fired from. For example, a round from an AK-47 assault rifle leaves the muzzle of the gun at twenty-three hundred feet per second, twice the speed of sound. When it hits a person, the density of the tissue forces the round to yaw to one side until it is traveling sideways or even backward. Shock waves ripple through the tissue and create a cavity that can be as much as eleven times the size of the bullet. The cavity lasts only a few thousandths of a second, but the shock waves that created it can shred organs that the bullet never even touches. In head wounds the temporary cavity is particularly devastating because the skull—being rigid—can respond to the sudden deformation only by bursting. If the gun barrel is actually touching the victim, rapidly expanding gases inside the barrel get trapped in the wound and blow blood and tissue back out. It is safe to assume that some of the killers in Studenica walked away covered in the people they killed.

Even if the bodies are not recovered, though, a very good idea of what went on at the moment of death can be had by something called bloodstain pattern analysis. Drops of blood splash differently depending on the angle at which they strike a hard surface, and arcsine equations can be used to reconstruct the path that the blood took through the air. If there are more than two bloodstains, something like triangulation can be used to figure out—down to an area about the size of a grapefruit—where in space they all originated. "That's useful in saying, 'Well, the person was standing up when they were shot,' " says Grant Graham, a criminalist with the Armed Forces Institute of Pathology. " 'The person was on their knees. The person was lying down.' . . . There are all different types of bloodstain patterns—swipes, wipes, drips, arterial spurts, gushes—and you can reconstruct what happens in the crime scene as things move along. . . . It's a moving, flowing event."

Unfortunately, many of the dead in Kosovo were burned beyond recognition. The report for 157 Millosh Gillic Street states that FBI investigators went to a burned-out residence and found "the skeletal remains of an indeterminate number of victims." But even a charred pile of bones can contain enough evidence to identify the dead and the method of death.

"The human skeleton is a dynamic part of the body," says Dr. Bill Rodriguez, the forensic anthropologist attached to the FBI team. "It is constantly altered by activity. In a runner, changes occur in the bones of the leg; in a dockworker, changes occur in the upper torso. Bones are just like fingerprints . . . their structures are so unique that you can make a positive identification."

To identify the remains, the investigators need something to compare them with. Determining that you have a dead male in his mid-thirties who probably did a lot of heavy lifting doesn't do much good unless you also have descriptions of missing people to pick from. X rays provide the most accurate matches, as do dental work and injuries, but absent those kinds of records, forensic anthropologists can still provide a huge amount of information about who was killed and how. There is only one right femur in a body, for example; the number of right femurs in a pile of bones tells you how many people there were. Not only that, but bones differ enough between sexes, races, and age groups that it is often possible to define quite narrowly who these people must have been.

In 1948 an anthropologist working for the army's Office of the Quartermaster General—using war dead from Guadalcanal, Iwo Jima, and other Pacific battlegrounds—improved upon an already strong statistical connection between body height and femur length. If you multiply the length of the femur by 2.38 and then add 61.41 centimeters, you get the height of a person to within a fraction of a centimeter. Further military studies found that the shafts of the long bones of the body, known as the diaphyses, gradually form solid bonds with the caps of the bones throughout a person's teens. Different bones solidify at different times, providing a good indication of the dead person's age. Hundreds of dead American soldiers were identified using medical records and such statistical techniques; the same techniques can be applied to determine whether the dead in Kosovo's mass graves match accounts given by people who witnessed the killings.

"From a war crimes standpoint, we document the number of individuals, determine the sex ratio, and then determine the ratio of adults to children," says Rodriguez. "From there, we can move on to identifying the dead individually. We can also find evidence of blunt force trauma, stab wounds, gunshot wounds, malnutrition, and torture. . . . We deal with death on a daily basis. We're scientists of death."

The FBI convoy rolled out of Studenica in midafternoon. The family had reconvened in front of the house and was going through a pile of clothes that had been pulled up from the well along with the bodies. There were ski parkas, blankets, sweaters, a fake-fur coat. One man spotted the fake fur, knelt on it, and started sobbing. Xhevat, the grandson who had just returned from Germany, stood around a little uncomfortably, unsure what to do.

"Tell me about him," I said to Xhevat, pointing to the grave where the old man was now reburied.

"He was a farmer. He saw World War One and World War Two," Xhevat said. "He was a soldier for the Germans in World War Two. He always lived right here; he never considered leaving. Everybody fled, but he didn't. 'You are on my land. . . . No one will come and throw us out of here,' he would say."

"How long are you staying?" I asked him.

"Oh," Xhevat said, glancing at the rotting clothes in his front yard. "I go back to Germany on Friday."

In February 1994 German police took into custody a Bosnian Serb named Dusko Tadić, who had been hiding in his brother's apartment in Munich. Tadić had been recognized by Bosnian Muslims who had survived the infamous Serb death camp of Omarska, near the city of Prijedor. Tadić wasn't officially in the Bosnian Serb Army—he was a local café owner and karate teacher—but he would show up in the evenings to direct personally the torture of chosen prisoners. An ICTY indictment one year later accused him of rape, torture, thirteen murders, and—in his most infamous act—forcing a prisoner to bite the testicle off another, who subsequently died. Two years later, after a trial that lasted from May to November 1996, he was convicted of violating the laws and customs of war and of crimes against humanity and was sentenced to twenty years in prison.

The Tadić case was the first successful prosecution of a war crime by the ICTY. It has indicted ninety people for crimes committed in the former Yugoslavia and has an unknown number of secret indictments. Thirty of the accused are now in custody, including ten Bosnian Croats who surrendered in 1997 and three Bosnian Serbs who surrendered in 1998. One was killed resisting arrest. On July 25, 1995, South African Judge Richard Goldstone, acting as the tribunal's chief prosecutor, indicted Bosnian Serb President Radovan Karadžić and his chief of staff, General Ratko Mladić, with genocide, war crimes, and crimes against humanity. Four years later, neither one has been caught; Mladić is said to spend his days on a Bosnian Serb military base, tending beehives and a herd of goats. Each goat has been named after a Western leader or UN commander.

Considering the time and effort that it takes to indict someone for war crimes, the question of whether the investigations serve a purpose—in the absence of arrests—is a hard one to avoid. Judge Goldstone, who now heads an independent commission on NATO's use of force in Kosovo, is adamant that they do. "To turn people for the rest of their lives into international pariahs is not something that any rational person would like," he says. "Karadžić was no longer able to continue in office and had to disappear from the international scene. . . . The quality of their life is diminished considerably. . . . One is looking over one's shoulder every day of one's life."

Slobodan Milošević, as president of Yugoslavia, is not looking over his shoulder in quite the same way, but his is a precarious existence nonetheless. There is widespread discontent in the army, the country's infrastructure and economy are in ruins, and Montenegro wants to abolish the Yugoslav Federation. Because of the international arrest warrants, Milošević cannot flee to another country, and the United States is withholding economic aid to Serbia until he has been removed from power. "It is in the interest of the peoples of Serbia that [he] be transferred to The Hague," maintains David Scheffer, United States ambassador-at-large for war crimes issues. "It will remain a very difficult proposition to bring Serbia into the international community and into, frankly, the New Europe if Serbia remains a de facto sanctuary for indicted war criminals. . . . We think the odds are with us."

Because of the turmoil in Yugoslavia, there is a good chance that Milošević will one day stand trial. (Indeed, he has reportedly started making inquiries about hiring an English lawyer.) He has already been charged with crimes against humanity and violations of the laws and customs of war and may be charged with genocide as well, the most serious breach of humanitarian law. To convict him, the ICTY would have to show that he intended to destroy, "in whole or in part," the Albanian population of Kosovo. The fact that his forces generally spared women and children does not disqualify him from charges of genocide; theoretically, even one murder could be considered genocide if there was intent to harm the rest of the group. By that standard, the policies of the Milošević government easily qualify.

The atrocities in Kosovo are so well documented that Milošević will probably not bother to challenge them in court; instead, he may try to claim that he was unaware they were happening. He may already have had an eye toward that kind of defense when he encouraged paramilitary and militia forces to carry out many of the actual massacres in Kosovo. NATO has been eavesdropping on Serbian forces in the field since the beginning of the air war, though, and that will make an ignorance defense difficult. "To prove chain of command, we are relying to a great extent on intelligence agencies in the Western countries," says Landale, spokesman for the ICTY. "They are obliged to give it to us. We will try to show knowledge of crimes and failure to prevent them or to punish them."

When one hears accounts of the massacres in Kosovo, one is struck by both their terrible efficiency and their even more terrible savagery. Many of those responsible were hastily deputized militiamen who, to judge by the sheer creativity they showed in killing people, must have been quite intoxicated by their sudden power. They shot people one by one, and they mowed them down in groups; they burned them alive, and they cut their throats; they tortured them, and they just walked up and shot them in the backs of the heads. They briefly wielded absolute power in their brutal little lives and must have never stopped to think that they might one day be held accountable for it.

THE TERROR OF SIERRA LEONE

2000

The sin of Judah is written with a pen of iron,
and with the point of a diamond.
—JEREMIAH 17:1

Josephus, the son of an up-country diamond trader, checked over his shoulder and pulled a pack of 555-brand cigarettes out of his pocket. He opened the top and shook two diamonds into his palm, a twenty-five-carat coffee-colored industrial and a three-carat white gem. They looked like rock candy. We were at the Cape Sierra Hotel, one of the few safe places in Freetown, and Josephus wanted to do business.

"How do I know they're real?" I asked.

Josephus picked up a beer glass and dragged a long scratch down the side with the white. Very few things are hard enough to scratch glass, and a diamond is one of them. Josephus said that his father was a local chief in Kono and had large mining concessions there. Kono, a district in the northeastern corner of the country, is the richest diamond-producing area in Sierra Leone and—not coincidentally—is still under rebel control. Every two weeks, Josephus said, he flew to

Freetown to sell diamonds and returned with rice and palm oil for the miners. The miners were paid a dollar a day, and if they found any stones, they got a commission.

I could sense the bartender watching us. Josephus slid the stones back into their hiding place and said he could get more if I was interested. I told him I had to think about it. I was in Sierra Leone to write about the diamond trade, but being taken for an investor was virtually unavoidable. No one believed for a second that my photographer, Teun Voeten, and I were just journalists; Sierra Leone has been run as a mining scheme for the last seventy years, and there was no reason we should see it any differently. Before we arrived, a London contact had set up a meeting for us with one of the most powerful men in the Sierra Leonean military. Not for an interview, which he never would have consented to, but for a diamond deal.

Of all days for business, though, this was a bad one. Word had just gotten out that UN peacekeepers had surrounded Foday Sankoh's house in retaliation for rebel attacks elsewhere in the country. Sankoh is head of the RUF—the Revolutionary United Front—as the rebels call themselves, and under the UN-sanctioned Lomé Peace Accord of July 1999 he was given a government position and a compound outside Freetown. The day before, his forces in the field—possibly without consulting him—had encircled a UN disarmament camp in the town of Makeni and demanded that the Kenyan peacekeepers turn over ten rebels who had voluntarily surrendered their weapons. The commander refused, shooting broke out, and seven UN personnel were taken prisoner. The rest were still surrounded.

The last time Sankoh was arrested, the government had sentenced him to death. In response, the RUF nearly overran Freetown.

I told Josephus I'd look him up in a few days, and then I paid the tab and walked out of the hotel. The sudden thick dusk of the tropics had just dropped, and I could see garbage fires winking on the hillside above town. I dodged the crowd of hookers in front of the hotel and got into a hired car and told the driver to take me to Sankoh's house. The driver hesitated and then said he'd have to double his usual rate. We drove out across Aberdeen Bridge and through the roadside markets and shantytowns of Lumley, on the outskirts. Sankoh's compound was on a hill overlooking town; it consisted of an ugly yellow villa with a wall around it and a gutted concrete structure that served as a bunker. We pulled up to a flimsy checkpoint in the driveway, and a single UN peacekeeper stepped forward and asked us what we wanted. There were no other peacekeeping troops, no white-painted UN vehicles; the place was deserted. Suddenly a dozen young toughs in street clothes came running out of the bunker.

"Who are you? What do you want?" they shouted, pushing the peacekeeper aside. I explained that I was a journalist and had come to talk to Sankoh, but that was clearly not the right answer. They screamed that he wasn't in, and one of them started pounding on the roof of the car.

"Turn around," I told the driver. "Get us out of here."

The driver threw a fast U-turn, and we raced back the way we came. Halfway down the hill we pulled over to make way for a convoy of pickup trucks filled with more of Sankoh's boys. They weren't armed, but they were singing and pumping their fists in the air, as if they knew something.

As it turned out, they did.

The RUF started quietly and brutally when a hundred or so lightly armed guerrillas crossed into Sierra Leone from war-torn Liberia in late March 1991. Their intent was to overthrow the one-party system of Joseph Saidu Momoh, but the force included a large number of mercenaries from Liberia and Burkina Faso, and the campaign quickly devolved into an excuse to loot and kill. Playing off traditional male initiation rites, the rebels abducted children and teenagers, took them into the bush, and tattooed them with identifying marks so that they couldn't return to society unnoticed.

The rebels' leader, Foday Saybana Sankoh, was a Sierra Leone Army corporal who had been jailed for seven years for his suspected involvement in a 1971 plot to overthrow the government. After getting out of prison, he set up a photography business in Kailahun District, on the Liberian border, and spent the next decade traveling around the diamond fields of eastern Sierra Leone. At first Sankoh claimed simply to want to rid the country of one-party rule, but his forces quickly distinguished themselves with a brutality that was exceptional even by the standards of African warfare. Their trademark was amputations, mostly of hands, as a tactic to terrorize the local population. It was one of the only uses of mass amputations in the history of warfare, and it gave the RUF, a small, poorly armed force that had no real backing, a power disproportionate to its size.

Announcing their attacks beforehand to inspire terror, the rebels swept through southern and eastern Sierra Leone in a matter of months. The national army was too small, too disorganized, and too corrupt to offer much resistance. Some of them even joined forces with the rebels to loot. By 1995 the rebels were on the outskirts of Freetown, and President Valentine Strasser, a twenty-nine-year-old army officer who himself had seized power three years earlier, hired the South African security firm Executive Outcomes to deal with the problem. Making great use of several MI-24 helicopter gunships, Executive Outcomes took only a matter of weeks to drive the rebels out of Freetown and then out of Kono—although they neglected to destroy every last rebel base. (That would later prove to have been a mistake.) The gunships reportedly were so effective that the rebels offered a seventy-five-thousand-dollar reward—payable in diamonds—to anyone who destroyed one.

Utterly beholden to Executive Outcomes, the country was reported to have given up huge mining concessions in the face of a bill equal to half its annual defense budget. (Executive Outcomes denied having received concessions.) By January 1996 Strasser had been replaced by Julius Maada Bio, who in turn was replaced by the current president, Ahmad Tejan Kabbah, in a democratic election. In many other countries this would have been the end of the story, but not in Sierra Leone. Disgruntled army officers who hadn't been paid in months ousted Kabbah in 1997, released six hundred inmates from Pademba Road Prison, brought the rebels into the government, and instituted their own brutal regime. They, in turn, were thrown out by ECOMOG, a Nigerian-led regional peacekeeping force that managed to reinstate Kabbah as president in early 1998. Kabbah, however, then made the mistake of executing twenty-four disloyal army officers and bringing Sankoh up on charges of treason. The charges stemmed from a 1997 arms-buying trip Sankoh had made to Nigeria on behalf of the RUF. The rebel leader was quickly found guilty, but before his death sentence could be carried out, a sketchy alliance of rebels and army irregulars staged another attack on Freetown.

War does not get much worse than January 6, 1999. Teenage soldiers, out of their minds on drugs, rounded up entire neighborhoods and machine-gunned them or burned them alive in their houses. They tracked down anyone whom they deemed to be an enemy— journalists, Nigerians, doctors who treated wounded civilians—and tortured and killed them. They killed people who refused to give them money, or people who didn't give enough money, or people who looked at them wrong. They raped women and killed nuns and abducted priests and drugged children to turn them into fighters. They favored Tupac T-shirts and fancy haircuts and spoke Krio—the common language of Freetown—to one another because they didn't share a tribal language. Some were mercenaries from Liberia and Burkina Faso, a few were white men thought to be from Ukraine, but most were just from the bush. They had been fighting since they were eight or nine, some of them, and sported such names as Colonel Bloodshed, Commander Cut Hands, Superman, Mr. Die, and Captain Backblast. They fought their way west in Freetown, neighborhood by neighborhood, through Calaba Town and Wellington and Kissy, and they weren't stopped until they had nearly overrun the ECOMOG headquarters at Wilberforce Barracks.

Eventually the Nigerian-led military machine set itself in motion. It attacked with heavy artillery and Alpha jets and helicopter gunships. Some of the gunships were piloted by white South Africans who just threw mortars out of the gun bays when they ran out of ammunition. Slowly the rebels fell back. Realizing that they were going to lose the city, they started rounding up people and detaining them until special amputation squads could arrive. The squads were made up of teenagers and even children, many of whom wore bandages where incisions had been made to pack cocaine under their skin. They did their work with rusty machetes and axes and seemed to choose their victims completely at random. "You, you, and you," they would say, picking people out of a line. There were stories of hands' being taken away in blood-soaked grain bags. There were stories of hands' being hung in trees. There were stories of hands' being eaten.

"They marched us at gunpoint to the hill near Kissy Mental," one fifteen-year-old girl named Ramatu later told human rights investigators. "They didn't say why they were taking us but we knew. . . . They had us get down on our knees and put our arms on a concrete slab. . . . One rebel did all the cutting. A few had both hands cut off; others just one. And then they walked away. I couldn't even bury my arm."

It took several weeks, but the Nigerians eventually drove the rebels out of Freetown and back up-country. Six thousand people had died in Freetown. Although the rebel assault had failed militarily, it had so traumatized the civilian population that it was prepared to do almost anything—including accept the rebels as part of the government—in order to bring an end to the war. The result was the Lomé Peace Accord, which granted a blanket amnesty to all combatants, instituted a nationwide disarmament program, opened the door to eleven thousand UN peacekeepers, and assigned government posts to rebel commanders. Sankoh was made vice-president of the country, as well as chairman of the Commission for the Management of Strategic Resources, National Reconstruction and Development.

That was a lot of words to say that he was now the diamond czar of Sierra Leone.

Everyone's fear—that the UN would surround Sankoh's house and arrest him—turned out to be unfounded, but the night I'd driven up there, the mood in the city was as tight as a piano wire. By dark the streets were empty, and around midnight bursts of automatic gunfire were heard in the hills outside Freetown. It turned out to be just skittish security forces shooting at one another. There were rumored to be thousands of RUF within the city itself, waiting for the signal to rise up, and no one knew when that moment would come. Teun and I were supposed to travel to the diamond fields up-country, and we were worried that if things got any worse, the planes would stop flying and we'd be stuck in Freetown. A contingent of rogue soldiers known as the Westside Boys had blocked the only road out of the city, and the UN was on the verge of suspending all internal flights because of the deteriorating security situation up-country. If Teun and I were going anywhere, we had to do it in a hurry.

The next morning we drove to a bullet-peppered airfield outside town and boarded an old twin prop that flew us up Bunce River and over the Moyamba Hills to the diamond-trading town of Bo, two hundred miles to the east. The first thing we did on the ground was check in with the commander of the Kamajors, a civil defense force made up of tribal hunters from the eastern part of the country. The Kamajors were wild fighters who terrified everybody, including the people they were defending, and until recently they had gone into battle wearing marine life jackets for effect. The Kamajors were supposed to be immune to bullets, and the rebels were so intimidated by Kamajor magic that in a sense it worked.

The commander assured us that God would take care of whatever the UN couldn't, which we took to mean that the Kamajors were busy rearming themselves, and then we wandered through town to talk to the diamond traders. Most of them had Lebanese names—Mansour, Jamil, Ahmad—and their offices were in small, brightly lit rooms tucked behind stores that sold radios and tools and dry goods and cloth—almost anything you'd want if you didn't want diamonds.

Teun and I were traveling with a longtime diamond miner named James Kokero, who had made and lost several small fortunes in Kono. His surname means "eagle," and among his associates he was known as the Eagle of Kono. Kokero, who was fifty, wore pressed shirts and slacks despite the heat and carried all his mining documents—twenty years' worth—in an old goat and snakeskin case. He said he had found his first diamond at age fifteen, when he stopped to relieve himself by the side of the road and realized he was pissing on a thirty-six-carat stone worth around twenty-eight thousand dollars. His father, who was already in the mining business, lost all the money from the sale of the stone on exploratory mining, so Kokero dropped out of school and wound up joining a gang called the Born Losers, which specialized in stealing gravel from the diamond fields. In Sierra Leone, gravel is money: Wash it, and sometimes there are diamonds in it. The Born Losers sold their gravel to Lebanese diamond traders who paid them a percentage of whatever stones turned up.

Kokero worked in the business off and on for the next twenty years, graduating to large foreign companies that invested hundreds of thousands of dollars in draglines and bulldozers for deep alluvial mining. Several times his operation was sabotaged, and his life was even threatened by Lebanese traders who were said to have had a very close relationship with the local authorities. When the war came, Kokero was working with an American named Mike Taylor up in Kono. One day a group of irregular soldiers seized their equipment and told the two miners that they were going to be killed. "Would you rather be shot or buried alive?" they asked. Taylor chose to be shot, so the soldiers stood them against a wall, and three men stepped up and cocked their machine guns. Kokero and Taylor both burst out laughing—it was all they could think to do—and this so puzzled their executioners that they demanded to know why they weren't scared.

"I'm a human being, like you," Kokero said. "We're brothers. If you kill me, you lose because you've killed a brother. For me, it's over, I'm gone. You're the one left with the problem."

The soldiers were so impressed with their fearlessness that they let the two men go. Kokero was a survivor, in other words, and our plan was to take him up to Kono and see if we could get a look at some of the illegal mining that the RUF was up to. The prospects looked bad, though. In Freetown we'd talked to an English photographer named Marcus Bleasdale, who was one of the few—and certainly the last—Western journalists to get into Kono. He and two Dutch reporters had driven through rebel roadblocks waving a letter from Sankoh himself, but when they arrived in Koido, the largest town in Kono, the local RUF commander told them straight out that the letter meant nothing. "Sankoh doesn't decide things here, I do," he said. He didn't let the reporters anywhere near the major diamond fields outside town, but small-scale mining was going on everywhere—along roads, behind mosques, anywhere they could find gravel. Locals would set up washing plants and sift through the gravel for diamonds; then the rebel command would come in and take its share.

It was the beginning of the rainy season, and the thunderstorms came in over Bo at the end of the afternoon: heavy towers of cumulus that turned the air yellow and rattled rain down so hard you couldn't see across the street. Men and women ducked under corrugated zinc awnings, and boys tore their shirts off and darted through the torrent like fish. At six-thirty the BBC came on the air and said that the UN had lost communication with some two hundred Zambian peacekeepers near Makeni, and that it was thought they had been surrounded and disarmed. Helicopter reconnaissance indicated that the RUF was now driving around in the Zambians' armored vehicles. "The rebels appear to be on the move," said UN spokesman Fred Eckhard on the broadcast. "But we don't know where."

———

Diamonds are not particularly rare geologically, and not particularly valuable intrinsically; they mainly cut things well, which makes them worth up to about thirty dollars a carat for most industrial applications. What gives diamonds tremendous economic power is the fact that 70 to 80 percent of the world's gem-quality diamonds flow through a group of companies collectively known as De Beers, which regulates the availability of diamonds so that prices remain high. In the late 1920s, when the diamond industry was in complete disarray, Sir Ernest Oppenheimer soaked up most of the world's supply and began price setting in such a way that the industry remained profitable. Today De Beers mines 50 percent of the nearly seven billion dollars' worth of the world's gem diamonds produced every year and buys another 20 to 30 percent through its Central Selling Organization. The CSO takes these diamonds, sorts them into shoebox-size parcels, and then sells them to a total of about 120 "sightholders" throughout the world. The sightholders often do not see the stones before they buy them and pay whatever price De Beers asks.

Approximately half the De Beers sightholders are based in Antwerp, Belgium, Europe's traditional diamond hub. Until recently a value added tax—a small fee levied on raw materials when they are processed—was so easy to dodge that a twenty-billion-dollar-a-year industry paid only eight million dollars a year in taxes. The industry is regulated by the Hoge Raad voor Diamant, the Belgian Diamond High Council, which serves both to represent Antwerp in the international market and to monitor the industry on behalf of the Belgian government. The council is charged with evaluating diamond imports and certifying their country of origin. For the purposes of the Diamond High Council, the country of origin is simply where the stone was last exported from. That clause—in a nutshell—is the heart of the illegal diamond trade.

Under the laws of Sierra Leone—which Sankoh was charged with upholding—every diamond mined in the country must be brought to a Government Gold and Diamond Office to be weighed, classified, and assigned a value. If the licensed exporter wants to sell the stone, he pays a 2.5 percent tax, and the stone or parcel of stones is sealed in a box and stamped. The box is not supposed to be opened again until it reaches its destination. Foreigners often team up with citizens of Sierra Leone who hold mining licenses and then make arrangements with landowners to mine their land in exchange for a portion—usually between a third and a half—of whatever diamonds are found.

One of the reasons the export tax on diamonds is so low is that to some degree, it is a voluntary tax. Diamonds are the most concentrated form of wealth in the world; millions of dollars' worth can fit into a pack of cigarettes. Diamonds are so small, so valuable, and so easy to conceal that if taxes on them rise above a certain level, overall revenue falls because people simply start smuggling. Some people hide the stones on their person and board a plane for Belgium; others transport them overland to Guinea or Liberia and sell them on the local black market. The places to hide a diamond are almost limitless. They are heated and dropped into tins of lard. They are sewn into the hems of skirts. They are encased in wax and taken as suppositories. They are swallowed, hidden under the tongue, burrowed into the navel, or slipped into an open wound that is then allowed to heal.

A rebel group such as the RUF would not bother to resort to any of those measures; it would simply smuggle them overland. Diamonds are carried out on foot over the maze of jungle paths that connect Sierra Leone to Liberia, or they are taken out by light airplane. Marcus Bleasdale said that when he was in Kono, he heard planes landing and taking off regularly, though he wasn't allowed anywhere near the airstrip. Once in Liberia—or Guinea, or Burkina Faso—the stones are passed off as domestic and shipped to the international markets of Antwerp and Tel Aviv. According to reports by the United States Geologic Survey, the total output from all of Liberia's diamond mines is only 100,000 to 150,000 carats a year, yet the Diamond High Council logged Liberian diamond imports averaging six million carats a year between 1994 and 1998 alone. It is no mystery where the discrepancy comes from, and the same problem exists in Angola, where UNITA rebels have sold around three billion dollars' worth of illegally mined diamonds to fund a war that to date has killed half a million people.

This has all come to light in the West in just the past few months, beginning with a report about RUF diamond mining by a nonprofit group called Partnership Africa Canada. That was followed by a report from Robert Fowler, Canada's ambassador to the UN. Both papers made it quite clear: If international diamond brokers made a concerted effort to avoid buying illicitly mined diamonds, groups such as UNITA and the RUF would have a much, much harder time bankrolling their wars. Since then, De Beers has urged punitive action against any dealers trafficking in so-called conflict diamonds. By mid-June the UN proposed a ban on the export of all Sierra Leonean diamonds that have not cleared customs in Freetown. And the European Union decided to halt foreign aid to Liberia because of Liberian president Charles Taylor's support of the RUF.

Nonetheless, selling illicit diamonds in Antwerp is still just a matter of a few phone calls. And so for the past ten years, Sierra Leonean diamonds have flowed unchecked across the porous border of Taylor's corrupt little country. Not surprisingly, Taylor was one of the original supporters of Sankoh back in 1991, when the first hundred RUF fighters crossed over the Mano River. Equally unsurprising, Sankoh's posting as head of the Commission for the Management of Strategic Resources—diamonds, essentially—did absolutely nothing to stem the flow.

The diamond fields start right outside Bo; you can see them alongside the road east to Kenema. They're just gravel pits carved out of the jungle, dotted with teenage boys in their underwear shoveling mud. We drove out there the following day with James Kokero, racing along one of the only good highways in the country, past mud-walled villages and upland farms hacked out of the bush. Some clearings were still smoking from the burnovers that precede planting season. "I used to farm," said Kokero sourly, "farm and mine. You mine for the money; you farm to eat."

The young miners were friendly, stopping their work to ask for cigarettes when we pulled over. They worked in shifts in the hammering sun, digging down into the diamond-bearing gravel and piling it up on the side to be sorted. Alluvial mining is not dramatic or dangerous or even costly; it just requires a lot of people digging. Larger operations use draglines and bulldozers to get through what is known as the overburden, but people interested in those kinds of investments have mostly disappeared from Sierra Leone.

Almost anyone, however, can set up a small-scale alluvial mining operation. The diggers are fed rice twice a day, paid a nominal amount of money, and given a share of whatever diamonds are found. The gravel gets shoveled out of steep-sided pits and then pumped into small steel washing plants that are run off a generator. There it is mechanically sorted for size, sluiced for gold, and then carted off to a secluded area—usually behind a rattan fence—to be picked through for diamonds. Typically, a third of the stones are turned over to the workers, a third are kept by the financial backers, and a third are given to the landowner. Obviously, it's a system full of opportunities to steal someone blind.

Sierra Leone was founded in 1787 as a colony for slaves freed by the British during the American Revolution. Diamonds were discovered there in 1930. Legend has it that, when word got around, the British started telling locals that the stones were electric and dangerous to touch. Their advice was to leave them alone until a white man could get there. On a larger scale, that was essentially how the colonial government of Sierra Leone handled its newfound wealth: In 1937 it sold a De Beers–owned company exclusive mining rights to the entire country for the next ninety-nine years. De Beers quickly got production levels up to a million carats a year, but it was only a matter of time before the locals realized that instead of working for De Beers they could just find diamonds on their own. Soon there were tens of thousands of illicit miners in Kono washing river gravel in homemade sieves and selling whatever they found to Lebanese and Mandingo traders. At first, the traders sold their stones in Freetown, but then, when that got too difficult, they smuggled them across the Mano River into Liberia.

By the 1950s, 20 percent of the stones on the world market were thought to have been smuggled out of Sierra Leone, mostly through Liberia. De Beers found itself facing a choice: Lose control altogether of the Sierra Leone diamond trade or open an office in Monrovia, the capital of Liberia, to buy back all the stones that were being mined illegally. Of course, they set up the buying office. In the end the licensing system proved untenable, and in 1963 the newly independent government of Sierra Leone bought back most of the mining rights to the country. For the first time, diamond licenses were made available to the locals, and a patronage system developed whereby diamond buyers—Lebanese, for the most part—fronted people money to start mining operations and then bought the stones that were found.

In the 1980s De Beers closed its buying office in Liberia, but that has done little to impede the flow of Sierra Leonean diamonds to Antwerp. Now the majority of people running mining operations up-country are local Lebanese and a handful of foreigners. We found Gregg Lyell drinking a Coke at the Capitol Bar in Kenema. Kokero, who seemed to know everybody, spotted him and brought him over. Lyell, now in his fifties, is an American who came to Sierra Leone several years ago to buy diamonds and wound up staying. He married a local woman and sat out the 1997 coup in Freetown with a gun on his lap. Now he was running a dredge mine that sucked gravel off the bottom of the Sewa River between Kenema and Bo.

"Dredge mining is all hit-or-miss," Lyell explained. "The divers take a propane bottle and an air compressor, stick a hose in it, tie a rag around their eyes to keep the dirt out, and go down and dredge. You pump everything into a canoe, drag it to shore, and go through it with a kicker"—a sieve—"and then flip that over on the bank. Diamonds are heavier than most other stones, so the ones that worked their way down to the bottom of the kicker will now be on top."

Dredging can be dangerous, but that's where the diamonds collect—in the gravel along the river bottom. There are supposed to be enormous diamond deposits off the coast, at the mouths of the Sewa and Mano rivers, but seabed dredging is extremely expensive. Lyell said his divers worked thirty to fifty feet down for half an hour at a time and wore sandbag weight belts to keep themselves on the river bottom. Some divers are known to sacrifice sheep before starting to work. They make sure the blood mixes with the river water to safeguard their lives.

"I started studying diamonds back in the States," Lyell said. "Let's just say that once upon a time I was a bad boy and found myself with a lot of time on my hands. . . . I'll probably stay here for a while. I was supposed to go to Mali to buy some gold, but that didn't happen."

Lyell wore his hair cropped short in front with a ponytail and had the beginnings of a thin goatee. Like everyone else, he was sweating heavily in the afternoon heat. A truck filled with miners rattled by at one point, and Lyell pointed at it. "Tongo Field," he said. "Trucks go up there every day."

"Tongo Field?" I asked. "Isn't that RUF territory?"

Lyell didn't say anything. He just looked at me with an expression that I'd already begun to recognize: the expression of someone who has devoted his entire life to diamonds and finds himself dealing with someone who hasn't.

By the time we left Kenema, three days later, the situation had deteriorated to the point where we'd begun to wonder if we'd even have trouble getting back into Freetown. As many as five hundred peacekeepers were now being held hostage around the country, a Guinean Army contingent had been forced to flee an important base called Rogberi Junction, and the rebels were rumored to have reached Hastings Airport, on the outskirts of Freetown. This last proved to be untrue, but just the rumors were enough to trigger widespread panic. It was starting to look like January 6 all over again.

There were checkpoints on the Bo–Kenema road every few miles now, and they were manned by Kamajors with guns. These were the first guns we'd seen in the country, apart from UN peacekeepers' weapons, and it was a bad sign; it meant that the government had given up on the UN and had decided to take matters into its own hands. As soon as we drove into Bo, it was clear something was up; there were too many groups of men on the street, too many trucks rumbling in and out of town. We dropped our bags off at the hotel and walked back to the Civil Defense Headquarters, where we'd seen a crowd of several hundred Kamajors.

The commotion started as soon as we arrived. *"We de go kill dem! We do go kill dem!"* one Kamajor started shouting in Krio, jamming a round into his grenade launcher. He climbed into a car with five or six others and sped off down the street. The weapons had materialized out of nowhere, and every man had one: rocket-propelled grenade launchers and Kalashnikovs and sleek black FN assault rifles and even old shotguns and sabers left over from colonial days. They had come from the bush, these men, and they'd brought with them their protective magic and their claims of special powers. They wore sackcloth tunics and fishnet shirts studded with crocheted pouches that were supposed to stop bullets. They sewed cowrie shells onto their clothing and wore bone necklaces that hung down over their ammo belts and clacked against their guns. One guy had nothing on but shorts and a pink ski parka hood. Another had a headband made of live machine-gun rounds. They stood in angry little clusters around shortwave radios listening to the afternoon BBC report and slapping ammo clips into their guns.

That morning, apparently, several thousand protesters had gathered at Sankoh's compound to protest the war, and Sankoh's bodyguards had opened fire. Television footage showed teenage boys in civilian clothes emptying banana clips into the crowd. One bodyguard even fired off a rocket-propelled grenade. Some accounts had Sankoh pleading with his bodyguards not to shoot, and other accounts had him standing on the balcony with a machine pistol, directing the attack. Nineteen civilians were killed, and scores were wounded. Later that day a group of irregulars stormed the house and killed some of the bodyguards, but Sankoh himself had fled. There were rumors that he had escaped in a UN vehicle, or that he was hiding in Freetown, or that he'd fled into the bush and was making his way back to rebel lines. Government forces rounded up two dozen RUF officials in Freetown and detained them, and Kamajors had done the same thing in Bo. In the meantime the rebels were advancing down the road to Freetown and had hit a town called Waterloo, only twenty miles away.

"You didn't hear it from me," a UN military observer near Bo told me that afternoon, "but it's going to be like the fall of Saigon when we pull out."

The next morning, British SAS in two big Chinook helicopters came pounding in low from the south and landed at the dirt airfield outside Bo. They took on twenty or thirty foreign-passport holders, including Teun and me, and then roared back to Freetown. They flew twenty feet above the forest canopy, and when we passed over little villages, we could see people run out of their huts to watch.

The first place Teun and I went when we got back to Freetown was Sankoh's house. It was early morning, and there was no one there; the gate had been torn off its hinges, and twisted clothes and spent bullets littered the yard. We stepped inside and sloshed through water that was three inches deep over the marble floors. Somewhere it was still running, gurgling out of a pipe where protesters had torn the plumbing out of the walls. There were women's panties and bras on the towel rack in the bathroom, as well as an empty bottle of 1998 Laurent Grand Siècle Ferme. In the upstairs bedroom there was an empty box of 70-mm ammo. Papers were scattered everywhere, and syringes—thousands of them, used and unused—lay piled in the corners like drifted snow.

Long before we'd gotten there, other journalists and Sierra Leonean detectives had scoured the premises for incriminating documents. According to Minister of Information Julius Spencer, they found evidence that Sankoh had organized a coup for Tuesday, May 9, but the protest at his compound the day before had derailed it. A number of rebel commanders, including Denis ("Superman") Mingo, Colonel Akim, and Brigadier Issa Sesay, and at least one Ukrainian mercenary had infiltrated the city to coordinate the uprising. Some of these men were killed or arrested in the days following the massacre. The bodyguards I'd seen driving up the hill to Sankoh's house, pumping their fists and singing: They all had been thinking that within days their leader would be in control of the capital. In that light, their bravado made perfect sense.

More important than evidence of a planned coup, however, were secret RUF reports on mining operations in Kono. A blue composition book appeared to list every diamond collected by just one RUF officer between October 30, 1998, and July 31, 1999. The book had been meant for use by schoolchildren and had "God Bless the Teacher" and "PEACE" printed on the cover. For NAME the owner had written in careful script, "Capt Joseph 'K' Bakundu." For SCHOOL he'd written, "R.U.F. Minning Unite." And for CLASS he'd written, "Black guard."

The Blackguards were Sankoh's elite bodyguard unit. Bakundu apparently was responsible for collecting diamonds from about fifteen rebel dealers up in Kono and Tongo Field, and they in turn had presumably collected them from diggers in the bush. Some of the names on the list—Sam Bockarie (known as Mosquito), Colonel Akim— were those of well-known rebel commanders. The book lists a nine-month haul of about 786 carats of white diamonds and 887 carats of industrials. The stones included a 17-carat orange, a 9-carat white, and numerous others between 1 and 6 carats. The RUF is thought to be exporting about half a million carats a year, which would suggest there were about three hundred guys like Bakundu gathering diamonds for Sankoh.

Not only was the RUF mining diamonds, but it was also in contact with Western businessmen. In his official capacity as chairman of the Strategic Resources commission, Sankoh had drawn up an agreement to buy and sell precious stones with Samuel Isidoor Weinberger of London. Sankoh had also negotiated with Raymond Clive Kramer of the Kramer Group of Companies in South Africa about expert consulting on mining operations. There was a letter from Patrick Everarts de Velp, the Walloon trade representative in Washington (Wallonia is part of Belgium), who was trying to arrange for the sale of some mining equipment to Sankoh. "It is always a great honour and a privilege to help you," de Velp wrote.

And there were many, many letters from an American named John Caldwell. Caldwell, the president of the U.S. Trading & Investment Company, in Washington, D.C., had tried to arrange agricultural deals through Sankoh, including a thirty-two-million-dollar food shipment. (Sankoh had opposed that particular deal because he didn't want the food to be handled by international relief organizations— presumably because they would not favor the RUF in their distribution.) Caldwell is a French-born naturalized American who served in NATO intelligence in the mid-1960s and then became vice-president of international affairs for the U.S. Chamber of Commerce. Last October, he and his business partner, a Belgian named Michel Desaedeleer, went to Freetown to negotiate what they say was a comprehensive development program for Sierra Leone. They claim that their idea was to bring in an international mining firm, such as De Beers, and use the revenue to fund agricultural projects in rural areas. In order to broker a deal of that magnitude, however, they needed to have something to offer, and last October 23, they got it. Sankoh signed a contract that gave them a monopoly on all gold and diamond mining in the rebel-controlled territory of Sierra Leone. The contract was between the RUF and the BECA Group, an offshore company registered in Tortola, British Virgin Islands, which listed Desaedeleer and Caldwell as directors. BECA was to run all mining operations in the RUF-controlled areas and handle all export and sale of diamonds on the international market. The RUF was to provide security and labor for the mining operations and facilitate the transportation of diamonds out of the country. BECA and the RUF would split all profits.

The contract specified that the agreement would become null and void as soon as the government of Sierra Leone activated the Commission for the Management of Strategic Resources, National Reconstruction and Development, of which Sankoh was chairman. At that point, a new contract would be negotiated between BECA and the commission. Until then, however, mining in Sierra Leone was wide open to anyone who wanted to do business with the RUF.

Upon returning to the United States, Desaedeleer went to the embassy of Sierra Leone and met with John Leigh, the Sierra Leonean ambassador to the United States. He showed Leigh the contract and offered to sell it to him for ten million dollars, which he claimed was its value on the open market. In effect, he was trying to sell something to the government of Sierra Leone that Sankoh had no legal basis for giving away in the first place. Not only did the RUF have no legal claim to mining rights in Sierra Leone, but even in his capacity as chairman of the Strategic Resources commission, Sankoh did not have the authority to negotiate a contract by himself. At the very least, he needed the signatures of the other members of the commission, which he obviously did not have. Shocked at the proposal—and its price— Ambassador Leigh says he asked to make a photocopy of the document so that he could send it to his government. Desaedeleer refused, and Leigh asked him to leave the embassy immediately.

After that, Caldwell and Desaedeleer tried to sell the license to various mining companies—De Beers, DiamondWorks, Rex, Rio Tinto—but were turned down by all of them. Finally, Desaedeleer says, he got the ear of Charles Finkelstein, a member of a prominent Antwerp diamond family. Finkelstein later denied any professional involvement with Desaedeleer, but at the time Desaedeleer seemed to think he had found a partner. At the very least, he may have thought that Finkelstein's name would impress Sankoh.

"With Charles, we can BUY," Desaedeleer wrote to Sankoh on April 6. "Charles has the financial ability to do anything, a private jet from Belgium to Kono or to Monrovia or to Freetown or any other solution. . . . What we have to solve: How will you convince the people in charge in Kono to bring everything to you instead of 10% and [if] it is not possible how are you going to convince them to sell those 90% to us instead of keeping it or selling it to the Lebanese or whoever? . . . Foday what I'm saying is this, the money is finally on the table, you make sure that the merchandise is available one way or another and all of us will be ok."

Desaedeleer may have been vying with half a dozen other Western businessmen—all pursuing mining contracts—for Sankoh's attention. In a sense these men were not the problem; they were just trying to exploit one. The real problem was that Sankoh was presiding over a system in which all the diamonds of Kono were being diverted from Freetown and smuggled out of the country. According to Ambassador Leigh, other documents found at Sankoh's house corroborate this; one even specified that 10 percent of the Kono diamonds went to Sankoh, 10 percent to the rebel commander Sam Bockarie, and 30 percent was used to buy arms and ammunition. The rest went to Liberian president Charles Taylor.

Weapons were the key. Without them the rebels could not control the diamond-producing regions, and without diamonds the rebels could not buy weapons. And there was plenty of evidence that weapons were making it into Sierra Leone. The British press reported that shortly before the January 6 invasion, a forty-ton shipment of weapons from Bratislava, Slovakia, had been flown into rebel-held eastern Sierra Leone by two British transport companies. And according to the New York–based organization Human Rights Watch, in April 1999, the ECOMOG commander in Sierra Leone reported that sixty-eight tons of weapons—including Strela-3 surface-to-air missiles and Metis guided antitank missile systems—had been flown into Burkina Faso on a Ukrainian-registered transport plane. From there, ECOMOG claimed, they were loaded onto smaller planes and flown into RUF territory. The end user certificate stipulated that the weapons could not be exported to another country, but in the fast and loose world of international arms trading, that hardly mattered.

"The arms trade in Africa works through brokers," a Belgian arms-trafficking authority named Johan Peleman told me before I arrived in Freetown. "They usually have a former intelligence or military background, but at the same time they are businessmen—commodity traders, for instance. . . . A typical broker would be a Belgian based in a French hotel room supplying guns from, I don't know, Lithuania, to a country neighboring the conflict zone. Documents would all look perfectly legitimate, but the arms end up with a rebel movement."

A couple of days before leaving Sierra Leone, we drove out to the front. The taxi driver wouldn't go beyond the town of Waterloo, so we got out and waited at a Nigerian Army checkpoint until a truckload of Kamajors drove up. They were headed twenty miles up the road to Masiaka, where a big battle had just taken place. They pulled us on board and veered back onto the road. There were about twenty of them, leaning against the sides of the truck and passing a joint around while the jungle blurred by on either side. At the deserted towns, soldiers who had been stranded would run out to try to wave us down, and going up through Occra Hills, we slowed to a crawl on the inclines while groups of Westside Boys watched us pass, pumping their guns in the air and screaming. From time to time we saw ambushed trucks with their engine parts sprayed out across the road, and around

Songo Junction there was the body of a rebel who'd been killed two days earlier. His corpse had turned foul so quickly on the hot asphalt that no one had bothered to drag him off.

Masiaka was at a crossroads that controlled access to the entire rest of the country; without it Freetown was basically under siege, and the rebels had held it for the past several days. But the Westside Boys had driven them out just hours earlier, and when we arrived, they were cranked out of their minds, either on coke or on the battle itself, and were milling around the town square, shooting their guns off. The Kamajors clambered down and joined in the shooting. Some government soldiers walked up, and within minutes an argument had broken out: something about who was doing the real fighting around here. An officer in the government forces began dressing down a Kamajor commander, and the Kamajor suddenly backed up a few steps and cocked his machine gun. The officer cocked his gun, and the Kamajors started cocking theirs, and suddenly everyone in the town square was screaming.

I glanced around for some cover, but all I could find was a concrete culvert along the road. We edged away and climbed into a pickup truck with some government soldiers. The rebels were in the bush a few miles away and a gun battle between Kamajors and government soldiers wasn't even close to being out of the question; it was time to get out of there. We drove back through the destroyed towns of Magbuntoso and Jama and then past the Nigerian forward positions and the Jordanian defenses around the airfield. Freetown was crowded and loud, the markets thronged with people and the streets completely choked by traffic. A British warship was visible out in the harbor. British paratroopers had dug bunkers into the hillside next to Aberdeen Bridge.

Africa stopped at Aberdeen. Europe began. We sat down at the terrace of the Mammy Yoko Hotel and ordered cold beers while the sun set and off-duty soldiers swam laps in the pool. Within a day we were clearing customs in Conakry and boarding an overnight flight to

Belgium. Sankoh was caught, in the end—spotted by an alert neighbor as he tried to sneak back into his house. Although the RUF released all the original UN hostages, they took more in June. Two foreign journalists, Kurt Schork of Reuters and Miguel Gil Morena de Mora of the Associated Press, were killed by rebels in a roadside ambush near Rogberi Junction. The rebels attacked Bo and Kenema and then withdrew to where they'd been three weeks earlier. The war continued up-country, although accounts of it rarely made it into the international press.

Very little had changed, really. Except that a few more people were dead.

THE LION IN WINTER

2001

The fighters were down by the river, getting ready to cross over, and we drove out there in the late afternoon to see them off. We parked our truck behind a mud wall, where it was out of sight, and then walked one by one down to the position. In an hour or so, it would be dark, and they'd go over. Some were loading up an old Soviet truck with crates of ammunition, and some were cleaning their rifles, and some were just standing in loose bunches behind the trees, where the enemy couldn't see them. They were wearing old snow parkas and blankets thrown over their shoulders, and some had old Soviet Army pants, and others didn't have any shoes. They drew themselves into an uneven line when we walked up, and they stood there with their Kalashnikovs and their RPGs cradled in their arms, smiling shyly.

Across the floodplain, low, grassy hills turned purple as the sun sank behind them, and those were the hills these men were going to attack. They were fighting for Ahmad Shah Massoud—genius guerrilla leader, last hope of the shattered Afghan government—and all along those hills were trenches filled with Taliban soldiers. The Taliban had grown out of the madrasahs, or religious schools, that had sprung up in Pakistan during the Soviet invasion, and they had emerged in 1994 as Afghanistan sank into anarchy following the Soviet withdrawal. Armed and trained by Pakistan and driven by moral principles so extreme that many Muslims feel they can only be described as a perversion of Islam, the Taliban quickly overran most of the country and imposed their ironfisted version of koranic law. Adulterers faced stoning; women's rights became nonexistent. Only Pakistan, Saudi Arabia, and the United Arab Emirates recognize their government as legitimate, but it is generally thought that the rest of the world will have to follow suit if the Taliban complete their takeover of the country. The only thing that still stands in their way are the last-ditch defenses of Ahmad Shah Massoud.

The sun set, and the valley edged into darkness. It was a clear, cold November night, and we could see artillery rounds flashing against the ridgeline in the distance. Hundreds of Taliban soldiers were dug in up there, waiting to be attacked, and hundreds of Massoud's soldiers were down here along the Kowkcheh River, waiting to attack them. In a few hours, they would cross the river by truck and make their way through the fields and destroyed villages of no-man's-land. Then it would begin.

We wished Massoud's men well and walked back to the truck. The stars had come out, and the only sound was of dogs baying in the distance. Then the whole front line, from the Tajik border to Farkhar Gorge, rumbled to life.

I'd wanted to meet Massoud for years, ever since I'd first heard of his remarkable defense of Afghanistan against the Soviets in the 1980s. A brilliant strategist and an uncompromising fighter, Massoud had been the bane of the Soviet Army's existence and had been largely responsible for finally driving them out of the country. He was fiercely independent, accepting little, if any, direction from Pakistan, which controlled the flow of American arms to the mujahidin. His independence made it impossible for the CIA to trust him, but agency officials grudgingly admitted that he was an almost mythological figure among many Afghans. He was a native of the Panjshir Valley, north of Kabul, the third of six sons born to an ethnic Tajik army officer. In 1974, he went to college to study engineering, but he dropped out in his first year to join a student resistance movement. After a crackdown on dissidents, Massoud fled to Pakistan, where he underwent military training. By 1979, when the Soviet Union invaded Afghanistan to prop up the teetering Communist government, Massoud had already collected a small band of resistance fighters in the Panjshir Valley.

As a guerrilla base the Panjshir couldn't have been better. Protected by the mountain ranges of the Hindu Kush and blocked at the entrance by a narrow gorge named Dalan Sang, the seventy-mile-long valley was the perfect staging area for raids against a highway that supplied the Soviet bases around Kabul, Afghanistan's capital. Massoud quickly organized his Panjshiri fighters, rumored to number as few as three thousand men, into defense groups comprising four or five villages each. The groups were self-sufficient and could call in mobile units if they were threatened with being overrun. Whenever a Soviet convoy rumbled up the highway, the mujahidin would mine the road, then wait in ambush. Most of the fighters would provide covering fire while a few insanely brave men worked their way in close to the convoy and tried to take out the first and last vehicles with rocket-propelled grenades. With the convoy pinned down, the rest of the unit would pepper it with gunfire and then retreat. They rarely stood and fought, and the Soviets rarely pursued them beyond the protection of their armored vehicles. It was classic guerrilla warfare, and if anything, Massoud was amazed at how easy it was. For his defense of the valley, Massoud became known as the Lion of Panjshir.

Very quickly, the Soviets understood that there was no way to control Afghanistan without controlling the Panjshir Valley, and they started attacking it with forces of up to fifteen thousand men, backed by tanks, artillery, and massive air support. Massoud knew that he couldn't stop them, and he didn't even try. He would evacuate as many civilians as possible and then retreat to the surrounding peaks of the Hindu Kush; when the Soviets entered the Panjshir, they would find it completely deserted. That was when the real fighting began.

Massoud and his men slept in caves and prayed to Allah and lived on nothing but bread and dried mulberries; they killed Russians with guns taken from other dead Russians and they fought and fought and fought, until the Soviets simply couldn't afford to fight anymore. Then the Soviets would pull back, and the whole cycle would start all over again.

Between 1979 and their withdrawal ten years later, the Soviets launched nine major offensives into the Panjshir Valley. They never took it. They tried assassinating Massoud, but his intelligence network always warned him in time. They made local peace deals, but he used the respite to organize resistance elsewhere in the country. The ultimate Soviet humiliation came in the mid-eighties, after the Red Army had lost hundreds of soldiers trying to take the Panjshir. The mujahidin had shot down a Soviet helicopter, and some resourceful Panjshiri mechanic patched it up, put a truck engine in it, and started running it up and down the valley as a bus. The Soviets got wind of this, and the next time their troops invaded, the commanders decided to inspect the helicopter. The last thing they must have seen was a flash; Massoud's men had booby-trapped it with explosives.

The night attack on the Taliban positions began with waves of Katyusha rockets streaming from Massoud's positions and arcing across the valley. The rockets were fired in volleys of ten or twelve, and we could see the red glare of their engines wobble through the darkness and then wink out one by one as they found their trajectories and headed for their targets. Occasionally an incoming round would explode somewhere down the line with a sound like a huge oak door slamming shut. The artillery exchange lasted an hour, and then the ground assault started, Massoud's men moving under the cover of darkness through minefields and machine-gun fire toward the Taliban trenches. The fighting was three or four miles away and came to us only as a soft, frantic *pap-pap-pap* across the valley.

We had driven to a hilltop command post to watch the attack. The position had a code name, Darya, which means "river" in Dari, the Persian dialect that's Afghanistan's lingua franca, and on the radio we could hear field commanders yelling, "Darya! Darya! Darya!" as they called in reports or shouted for artillery. The commander of the position, a gentle-looking man in his thirties named Harun, was dressed for war in corduroy pants and a cardigan. He was responsible for all the artillery on the front line; we found him in a bunker, studying maps by the light of a kerosene lantern. He was using a schoolboy's plastic protractor to figure out trajectory angles for his tanks.

Harun was working three radios and consulting the map continually. After a while a soldier brought in tea, and we sat cross-legged on the floor and drank it. Calls kept streaming in on the radios. "We've just captured another position; it's got a big ammo depot," one commander shouted. Another reported, "The enemy has no morale at all; they're just running away. We've just taken ten more prisoners."

Harun showed us on the map what was happening. As we spoke, Massoud's men were taking small positions around the ridgeline and moving into the hills on either side of a town called Khvajeh Ghar, which was at a critical part of the front line. Khvajeh Ghar was held by Pakistani and Arab volunteers, part of an odd assortment of foreigners—Burmese, Chinese, Chechens, Algerians—who are fighting alongside the Taliban to spread fundamentalist Islam throughout Central Asia. Their presence here is partly due to Saudi extremist Osama bin Laden, who has been harbored by the Taliban since 1997 and is said to repay his hosts with millions of dollars and thousands of holy warriors. The biggest supporter of the Taliban, however, is Pakistan, which has sent commandos, military advisers, and regular army troops. More than a hundred Pakistani prisoners of war sit in Massoud's jails; most of them—like the Taliban—are ethnic Pashtuns who trained in the madrasahs.

None of the help was doing the Taliban fighters much good at the moment, though. Harun switched his radio to a Taliban frequency and tilted it toward us. They were being overrun, and the panic in their voices was unmistakable. One commander screamed that he was almost out of ammunition; another started insulting the fighters at a neighboring position. "Are you crazy are you crazy are you crazy?" he demanded. "They've already taken a hundred prisoners! Do you want to be taken prisoner as well?" He went on to accuse them all of sodomy.

Harun shook his head incredulously. "They are supposed to represent true Islam," he said. "Do you see how they talk?"

I went into Afghanistan with Iranian-born photographer Reza Deghati, who knew Massoud well from several long trips he'd taken into the country during the Soviet occupation. Back then, the only way in was to take a one- to three-month trek over the Hindu Kush on foot, avoiding minefields and Russian helicopters, and every time Reza did it he lost twenty or thirty pounds. The conditions are vastly easier now but still unpredictable. Last summer, in a desperate effort to force international recognition for their regime, the Taliban launched a six-month offensive that was supposed to be the coup de grace for Massoud. Some fifteen thousand Taliban fighters—heavily reinforced, according to Massoud's intelligence network, by Pakistani Army units—bypassed the impregnable Panjshir Valley and drove straight north toward the border of Tajikistan. Their goal was to move eastward along the border until Massoud was completely surrounded and then starve him out. They almost succeeded. Waiting to go into Afghanistan that September and October, Reza and I watched one town after another fall into Taliban hands, until even Massoud's old friends began to wonder if he wasn't through. "It may be his last season hunting," as one journalist put it.

Massoud finally stopped the Taliban at the Kowkcheh River, but by then the season was so far advanced that the mountain roads were snowbound, and the only way for Reza and me to get in was by helicopter from the Tajik capital of Dushanbe. Massoud's forces owned half a dozen aging Russian military helicopters, and the Afghan embassy in Dushanbe could put you on a flight that left at a moment's notice, whenever the weather cleared over the mountains. On November 15, late in the afternoon, Reza and I got the word. We raced to the airfield, and two hours later we were in Afghanistan.

The helicopters flew to a small town just across the border called Khvajeh Baha od Din, and we were provided a floor to sleep on in the home of a former mujahidin commander who was now a local judge. Each night, anywhere from ten to twenty fighters stayed there, sleeping in rows on the floor next to us. The electricity was supplied by a homemade waterwheel that had been geared to a generator through an old truck transmission. Some fuel came in by truck over the mountains—a five-day trip—but farther north it all came in by donkey and cost twenty dollars a gallon. (The locals jokingly refer to donkeys as "Afghan motorcycles.") We washed at an outdoor spring and subsisted on rice and mutton and kept warm at night around a woodstove; we lived comfortably enough. The situation around us, though, was unspeakable.

Eighty thousand civilians had fled the recent fighting, adding to the hundred thousand or so who were already displaced in the north, and thousands of them were subsisting in a makeshift refugee camp along the Kowkcheh River half a mile away. They slept under tattered blue UN tarps and had so little food that some were reduced to eating grass. Tribal politics have long dominated Afghanistan; many observers, in fact, say that Massoud, a Tajik, will never be able to unite the country. These refugees were mostly ethnic Tajiks and Uzbeks, and they claimed that when the Taliban, who are Pashtun, took over a town, they raped the women, killed the men, and sold the young into servitude. One old man at a refugee camp pulled back his quilted coat to show me a six-inch scar on his stomach. A Taliban soldier, he said, had stabbed him with a bayonet and left him for dead.

A week or so after we arrived inside Afghanistan, Reza and I were told that Massoud was coming in—he'd been in Tajikistan, negotiating support from the government—and we rushed down to the river to meet him. A lopsided boat made of sheet metal, powered by a tractor engine that had paddle wheels instead of tires, churned across the Kowkcheh with Massoud in the bow. He wore khaki pants and Czech Army boots and a smart camouflage jacket over a V-neck sweater. He looked to be in his late forties and was as lean and spare as the photographs of him from the Soviet days. He was not tall, but he stood as if he were. The great man stepped onto the riverbank along with a dozen bodyguards and greeted us. Then we all drove off to the judge's compound in Khvajeh Baha od Din.

There he met with his commanders, listened to their preparations for the coming offensive, then hurried off. We later found out that he'd been forced to return to Tajikistan because of a chronically bad back; apparently the problem was so severe that it had put him in the hospital.

Finding ourselves once again waiting for Massoud, Reza and I decided to go out to the front line to see a position that had just been taken from the Taliban. We drove south along the Kowkcheh, past miles of trenches and bunkers, and stopped at an old Soviet base that had been gutted by artillery fire. The local commander was there, housed in the shell of the building. The wind whistled through the gaping windows, and his soldiers crouched in the shadows, preparing their weapons. The commander said that the position they'd taken was code-named Joy and that the bodies of the dead Taliban were still lying in the trenches.

He made a call on his radio and arranged for some men and pack-horses to meet us on the other side of the river. Then he directed us to the crossing point; it was in a canyon a few miles away, just below a town called Laleh Meydan. When we stopped there to sort our gear, a Taliban MiG jet appeared and made a pass over the town, completely ignoring the antiaircraft fire that was directed at it. The townsmen scattered but drifted back within minutes to help us carry our gear down to the river. The raft that was to ferry us across was made from a design that must have been around since Alexander the Great: eight cowhides sewn shut and inflated like tires, each stoppered by a wood plug in one leg and lashed to a frame made of tree limbs. Four old men paddled it across the river and then tied our gear to some horses. Three soldiers with Kalashnikovs were waiting to take us to the front.

It took us all afternoon to get there, walking and riding through mud hills, bare and smooth as velvet, that undulated south toward the Hindu Kush. There was no sound but the wind—not even any fighting—and nothing to look at but the hills and the great, empty sky. When we turned the last ridgeline, we saw Massoud's men silhouetted on a hilltop, waving us on.

Maybe the Taliban spotted our horses, or maybe they'd overheard the radio communications, but we were halfway up the last slope when I found myself facedown in the dirt as a Taliban rocket slammed into the hillside behind us. Then we were up and running, and the next rocket hit just as we got to the top, and they continued to come in, slightly off target, as we crouched in the safety of the trenches.

There was nothing exciting about it, nothing even abstractly interesting. It was purely, exclusively bad. Whenever the Taliban fired off another salvo, a spotter on a nearby hilltop would radio our position to say that more were on the way. The commander would shout a warning, and the fighters would pull us down into the foxholes, and then we'd wait five or ten seconds until we heard the last, awful whistling sound right before they hit. In a foxhole you're safe unless the shell drops right in there with you, in which case you'd never know it; you'd simply cease to exist. No matter how small the odds were, the idea that I could go straight from life to nonexistence was almost unbearable; it turned each ten-second wait into a bizarre exercise in existentialism. Bravery—the usual alternative to fear—also held no appeal, because bravery could get you killed. It had become very simple: It was their war, their problem, and I didn't want any part of it. I just wanted off the hill.

The problem was, "off" meant rising out of this good Afghan dirt we'd become part of and running back the way we'd come. Four hundred yards away was a hilltop that they weren't shelling; over there it was just another normal, sunny day. After we'd spent half an hour ducking the shells, the commander said he'd just received word that Taliban troops were preparing to attack the position, and it might be better if we weren't around for it. Like it or not, we had to leave. Reza and I waited for a quiet spell and then climbed out of the trenches, took a deep breath, and started off down the hill.

Mainly there was the sound of my breathing: a deep, desperate rasp that ruled out any chance of hearing the rockets come in. The commander stood on the hilltop as we left, shouting good-bye and waving us away from a minefield that lay on one side of the slope. Ten minutes later it was over: We sat behind the next ridge and watched Taliban rockets continue to pound the hill, each one raising a little puff of smoke, followed by a muffled explosion. From that distance, they didn't look like much; they almost looked like the kind of explosions you could imagine yourself acting bravely in.

The Taliban kept up the shelling for the next twelve hours and then attacked at dawn. Massoud's men fought them off with no casualties.

Massoud returned one week later, flying in by helicopter to Harun's command post to start planning a heavy offensive across the entire northern front. The post was at the top of a steep, grassy hill in some broken country south of a frontline town called Dasht-e Qaleh. It was late afternoon by the time we arrived, and Massoud was studying the Taliban positions through a pair of massive military binoculars on a tripod. The deposed Afghan government's foreign minister, a slight, serious man named Dr. Abdullah, walked up to greet us as we got out of the truck. Reza wished him a good evening.

"Good morning," Dr. Abdullah corrected, nodding toward the Taliban positions across the valley. "Our day is just beginning."

The shelling had started again, an arrhythmic thumping in the distance that suggested nothing of the terror it can produce up close. That morning, I'd awakened from a dream in which an airplane was dropping bombs on me, and in the dream I'd thrown myself on the ground and watched one of the bombs bounce past me toward a picnicking family. "Good," I'd thought; "it will kill them and not me." It was an ugly, ungenerous dream that left me unsettled all day.

Massoud knew where the Taliban positions were, and they obviously knew where his were, and the upshot was that you were never entirely safe. A guy in town had just had both legs torn off by a single, random shell. You couldn't let yourself start thinking about it or you'd never stop.

Massoud was still at the binoculars. He had a face like a hatchet. Four deep lines cut across his forehead, and his almond-shaped eyes were so thickly lashed that it almost looked as if he were wearing eyeliner. When someone spoke, he swiveled his head around and affixed the speaker with a gaze so penetrating it occasionally made the recipient stutter. When he asked a question, it was very specific, and he listened to every word of the answer. He stood out not so much because he was handsome but simply because he was hard to stop looking at.

I asked Dr. Abdullah how Massoud's back was doing. Dr. Abdullah spoke low so that Massoud couldn't hear him. "He says it's better, but I know it's not," he said. "I can see by the way he walks. He needs at least a month's rest . . . but, of course, that won't be possible."

The shelling got heavier, and the sun set, and Massoud and his bodyguards and generals lined up on top of the bunker to pray. The prayer went on for a long time, the men standing, kneeling, prostrating themselves, standing again, their hands spread toward the sky to accept Allah. Islam is an extraordinarily tolerant religion—more so than Christianity, in some ways—but it is also strangely pragmatic. Turning the other cheek is not a virtue. The prophet Muhammad, after receiving the first revelations of the Koran in A.D. 610, was forced into war against the corrupt Quraysh rulers of Mecca, who persecuted him for trying to make Arab society more egalitarian and to unite it under one god. Outnumbered three to one, his fighters defeated the Quraysh in 627 at the Battle of the Trench, outside Medina. Three years later he marched ten thousand men into Mecca and established the reign of Islam. Muhammad was born during an era of brutal tribal warfare, and he would have been useless to humanity as a visionary and a man of peace if he had not also known how to fight.

It was cold and almost completely dark when the prayers were finished. Massoud abruptly stood, folded his prayer cloth, and strode into the bunker, attended by Dr. Abdullah and a few commanders. We followed and joined them on the floor. A soldier brought in a pot for us to wash our hands, then spread platters of rice and mutton on a blanket. Massoud asked Dr. Abdullah for a pen, and Dr. Abdullah drew one out of his tailored cashmere jacket.

"I recognize that pen, it's mine," Massoud said. He was joking.

"Well, in a sense everything we have is yours," Dr. Abdullah replied.

"Don't change the topic. Right now I'm talking about this pen." Massoud wagged his finger at Dr. Abdullah, then turned to the serious business of preparing the offensive.

Massoud's strategy was simple and exploited the fact that no matter how one looked at it, he was losing the war. After five years of fighting, the Taliban had fractured his alliance and cut its territory in half. Massoud was confined to the mountainous northeast, which, although easily defensible, depended on long, tortuous supply lines to Tajikistan. The Russians, ironically, had begun supplying Massoud with arms—with the Taliban near their borders, they couldn't afford to hold a grudge—and India and Iran were helping as well. It all had to go through Tajikistan. The most serious threat to Massoud's supply lines came last fall, when he lost a strategically important town called Taloqan, just west of the Kowkcheh River. The Taliban, convinced that recapturing Taloqan was of supreme importance to Massoud, shipped the bulk of their forces over to the Taloqan front. Massoud arrayed his forces in a huge V around the town and began a series of focused, stabbing attacks, usually at night, that guaranteed that the Taliban would remain convinced that he would do anything to retake the town.

In the meantime he was thinking on a completely different scale. Massoud had been fighting for twenty-one years, longer than most of the Taliban conscripts had been alive. In that context, Taloqan didn't matter, the next six months didn't matter. All that mattered was that the Afghan resistance survive long enough for the Taliban to implode on their own. The trump card of any resistance movement is that it doesn't have to win; the guerrillas just have to stay in the hills until the invaders lose their will to fight. The Afghans fought off the British three times and the Soviets once, and now Massoud was five years into a war that Pakistan could not support forever. Moreover, the civilian population in Taliban-controlled areas had started to bridle under the conscription of soldiers and the harshness of Taliban law. Last summer, in fact, a full-fledged revolt boiled over in a town called Musa Qaleh, and the Taliban had to send in six hundred troops to crush it. "Every day, I bathe in the river without my pistol," the local Taliban governor later told a reporter, with no apparent irony. "What better proof is there that the people love us?" The end of the Taliban, it seemed, was only a matter of time.

The Dari word for war is *jang,* and as Massoud ate his mutton, he explained to his commanders that within weeks he would start a *jang-e-gerilla-yee.* Here in the north he was locked into a frontline war that neither side could win, but he had groups of fighters everywhere— even deep in areas the Taliban thought they controlled. "In the coming days, we will engage the Taliban all over Afghanistan," he announced. "Pakistan brought us conventional war; I'm preparing a guerrilla war. It will start in a few weeks from now, even a few days."

Massoud had done the same thing to the Soviets. In 1985 he had disappeared into the mountains for three months to train 120 commandos and had sent each of them out across Afghanistan to train 100 more. These 12,000 men would attack the vital supply routes of the cumbersome Soviet Army. They used an operations map that had been found in a downed Soviet helicopter, and they took their orders from Massoud, who had informers throughout the Soviet military, even up to staff general. All across Afghanistan, Russian soldiers traded their weapons for drugs and food. Morale was so bad that there were gun battles breaking out among the Soviet soldiers themselves.

Dinner finished, Massoud spread the map out on the floor and bent over it, plotting routes and firing questions at his commanders. He wanted to know how many tanks they had, how many missile launchers, how much artillery. He wanted to know where the weapons were and whether their positions had been changed according to his orders. He occasionally interrupted his planning to deliver impromptu lectures, his elegant hands slicing the air for emphasis or a single finger shaking in the harsh light of the kerosene lantern. His commanders—many of them older than he, most veterans of the Soviet war—listened in slightly chastised silence, like schoolboys who hadn't done their homework.

"The type of operation you have planned for tonight might not be so successful, but that's okay; it should continue," he said. "This is not our main target. We're just trying to get them to bring reinforcements so they take casualties. The main thrust will be elsewhere."

Massoud was so far ahead of his commanders that at times he seemed unable to decide whether to explain his thinking or to just give them orders and hope for the best. The Soviets, having lost as many as fifteen thousand men in Afghanistan, reportedly now study his tactics in the military academies. And here he was, two decades later, still waging war from some bunker, still trying to get his commanders to grasp the logic of what he was doing.

It was getting late, but Massoud wasn't even close to being finished. He has been known to work for thirty-six hours straight, sleeping for two or three minutes at a time. There was work to do, and his men might die if it wasn't done well, and so he sat poring over an old Soviet map, coaxing secrets from it that the Taliban might have missed. At one point he turned to one of the young commanders and asked him whether he could fix the hulk of a tank that sat rusting on a nearby hill.

"I have already been up there to see it," the young man said. "I have fixed tanks much worse than that."

There were a total of three destroyed tanks; Massoud thought they all could be salvaged. One was stuck in an alleyway between two houses, and the young commander said the passageway was too narrow for them to drag it out. "Buy the houses, destroy them, and get it out," Massoud said. "Get two more tanks from Rostaq; that's five. Paint them like new and show them on the streets so people will see them. Then the Taliban will think we're getting help from another country."

On and on it went, commander by commander, detail by detail. *Don't shell from Ay Khanom Hill; you're just wasting your ammunition. Don't shell any positions near houses or towns; the Taliban are too deeply dug in in those spots and you'll just hurt civilians. Send your men forward in jeeps to save the heavy machinery and shell heavily beforehand to raise a lot of dust. That way, the Taliban won't see the attack.*

When Massoud was growing up in Kabul, he was part of a neighborhood gang that had regular battles with other gangs. One particularly large gang would occupy a hilltop near his house, and he and his friends would go out and challenge them. Naturally enough, Massoud was the leader. He would split his force, sending one half straight up the hill while the other half circled and attacked from the rear. It always worked. It still worked.

Massoud sat cross-legged on the floor, bent forward at the waist, methodically opening and eating pistachios. His head hung low and swung from side to side as he spoke. He had a slight tic that ran like a shiver up his back and into his shoulders. "Get me your best guys," he said, looking around. "I don't want hundreds. I want sixty of your best. Sixty from each commander. Tomorrow I want to launch the best possible war."

———

Like so many fundamentalist movements, the Taliban were born of war. After the Soviet Union invaded Afghanistan on December 27, 1979, it ultimately sent in eight armored divisions, two enhanced parachute battalions, hundreds of attack helicopters, and well over a hundred thousand men. What should have been the quick crushing of a backward country, however, turned into the worst Soviet defeat of the Cold War. The very weaknesses of the fledgling resistance movement—its lack of military bases, its paucity of weapons, its utterly fractured command structure—meant that the Soviets had no fixed military objectives to destroy. Fighting Afghans was like nailing jelly to a wall; in the end there was just a wall full of bent nails. Initially using nothing but old shotguns, flintlock rifles, and Lee-Enfield .303s left over from British colonial days, the mujahidin started attacking Soviet convoys and military bases all across Afghanistan. According to a CIA report at the time, the typical life span of a mujahidin RPG operator—rocket-propelled grenades were the antitank weapon of choice—was three weeks. It's not unreasonable to assume that every Afghan who took up arms against the Soviets fully expected to die.

Without the support of the villagers, however, the mujahidin would never have been able to defeat the Soviets. They would have had nothing to eat, nowhere to hide, no information network—none of the things a guerrilla army depends on. The Soviets knew this, of course, and by the end of the first year—increasingly frustrated by the stubborn mujahidin resistance—they turned the dim Cyclops eye of their military on the people themselves. They destroyed any village the mujahidin were spotted in. They carpet-bombed the Panjshir Valley. They cut down fruit trees, disrupted harvests, tortured villagers. They did whatever they could to drive a wedge between the people and the resistance. Still, it didn't work. After ten years of war, the Soviets finally pulled out of Afghanistan, leaving behind a country full of land mines and more than one million Afghan dead.

A country can't sustain that kind of damage and return to normal overnight. The same fierce tribalism that had defeated the Soviets—"radical local democracy," the CIA termed it—made it extremely hard for the various mujahidin factions to get along. (It would be three years before they would be able to take Kabul from the Communist regime that the Soviets had left.) Moreover, the mujahidin were armed to the teeth, thanks to a CIA program that had pumped three billion dollars' worth of weapons into the country during the war. Had the United States continued its support—building roads, repatriating the refugees, clearing the minefields—Afghanistan might have stood a chance of overcoming its natural ethnic factionalism. But the United States didn't. No sooner had the Soviet-backed government crumbled away than America's Cold War–born interest in Afghanistan virtually ceased. Inevitably, the Afghans fell out among themselves. And when they did, it was almost worse than the war that had just ended.

The weapons supplied by the United States to fight the Soviets had been distributed through Pakistan's infamous Inter-Services Intelligence branch. The ISI, as it is known, had chosen a rabidly anti-Western ideologue named Gulbuddin Hekmatyar to protect its strategic interests in Afghanistan, so of course the bulk of the weapons went to him. Now, using Hekmatyar to reach deep into Afghan politics, Pakistan systematically crippled any chance of a successful coalition government. As fighting flared around Kabul, Hekmatyar positioned himself in the hills south of the city and started raining rockets down on the rooftops. His strategy was to pound the various mujahidin factions into submission and gain control of the capital, but he succeeded merely in killing tens of thousands of civilians. Finally, in exchange for peace, he was given the post of prime minister. But his troops remained where they were, the barrels of their tanks still pointed down at the city they had largely destroyed.

While the commanders fought on, life in Afghanistan sank into a lawless hell. Warlords controlled the highways; opium and weapons

smuggling became the mainstay of the economy; private armies battled one another for control of a completely ruined land. This was one
of the few times that Massoud's forces are thought to have committed
outright atrocities, massacring several hundred to several thousand
people in the Afshar District of Kabul. There is no evidence, however,
that Massoud gave the orders or knew about it beforehand. As early
as 1994 Pakistan, dismayed by the fighting and increasingly convinced
that Hekmatyar was a losing proposition, began to look elsewhere for
allies. Its attention fell on the Taliban, who had been slowly gaining
power in the madrasahs while Afghanistan tore itself apart. The
Taliban were religious students, many of them Afghan refugees in
Pakistan, who were trained in an extremely conservative interpretation
of the Koran called Deobandism. Here, in the tens of thousands of
teenage boys who had been orphaned or displaced by the war, Pakistan
found its new champions.

Armed and directed by Pakistan and facing a completely fractured alliance, the Taliban rapidly fought their way across western
Afghanistan. The population was sick of war and looked to the Taliban
as saviors, which, in a sense, they were, but their brand of salvation
came at a tremendous price. They quickly imposed a form of Islam
that was so archaic and cruel that it shocked even the ultratraditional
Muslims of the countryside. With the Taliban closing in on Kabul,
Massoud found himself forced into alliances with men—such as
Hekmatyar and former Communist Abdul Rashid Dostum—who
until recently had been his mortal enemies. The coalition was a shaky
one and didn't stand a chance against the highly motivated Taliban
forces. After heavy fighting, Kabul finally fell to the Taliban in early
September 1996, and Massoud pulled his forces back to the Panjshir
Valley. With him were Burhanuddin Rabbani, who was the acting
president of the coalition government, and a shifty assortment of mujahidin commanders who became known as the Northern Alliance.
Technically, Rabbani and his ministers were the recognized government of Afghanistan—they still held a seat at the UN—but in real-

ity, all they controlled was the northern third of one of the poorest countries in the world.

Worse still, there was a growing movement from a variety of Western countries—particularly the United States—to overlook the Taliban's flaws and recognize them as the legitimate government of the country. There was thought to be as much as two hundred billion barrels of untapped oil reserves in Central Asia and similar amounts of natural gas. That made it one of the largest fossil fuel reserves in the world, and the easiest way to get it out was to build a pipeline across Afghanistan to Pakistan. However appalling Taliban rule might be, their cooperation was needed to build the pipeline. Within days of the Taliban takeover of Kabul, a U.S. State Department spokesman said that he could see "nothing objectionable" about the Taliban's version of Islamic law.

While Massoud and the Taliban fought each other to a standstill at the mouth of the Panjshir Valley, the American oil company Unocal hosted a Taliban delegation to explore the possibility of an oil deal.

The day before the offensive, Massoud decided to go to the front line for a close look at the Taliban. He couldn't tell if the attack, as planned, would succeed in taking their ridgetop positions; he was worried that his men would die in a frontal assault when they could just as easily slip around back. He had been watching the Taliban supply trucks through the binoculars and had determined that there was only one road leading to their forward positions. If his men could take that, the Taliban would have to withdraw.

Massoud goes everywhere quickly, and this time was no exception. He jumped up from a morning meeting with his commanders, stormed out to his white Land Cruiser, and drove off. His commanders and bodyguards scrambled into their own trucks to follow him.

The convoy drove through town, raising great plumes of dust, and then turned down toward the river and plowed through the braiding channels, muddy water up to their door handles. One truck stalled in midstream, but they got it going again and tore through no-man's-land while their tanks on Ay Khanom Hill shelled the Taliban to provide cover. They drove up to the forward positions and then got out of the trucks and continued on foot, creeping to within five hundred yards of the Taliban front line. This was the dead zone: Anything that moves gets shot. Dead zones are invariably quiet; there's no fighting, no human noise, just an absolute stillness that can be more frightening than the heaviest gunfire. Into this stillness, as Massoud studied the Taliban positions, a single gunshot rang out.

The bullet barely missed one of his commanders—he felt its wind as it passed—and came to a stop in the dirt between Massoud's feet. Massoud called in more artillery fire, and then he and his men quickly retraced their route to the trucks. The trip had served its purpose, though. Massoud had identified two dirt roads that split in front of the Taliban positions and circled behind them. And he had let himself be seen on the front line, reinforcing the Taliban assumption that this was the focus of his attack.

Late that afternoon Massoud and his commanders went back up to the command post. The artillery exchanges had started up again, and a new Ramadan moon hung delicately in the sunset over the Taliban positions. That night, in the bunker, Massoud gave his commanders their final instructions. The offensive was to be carried out by eight groups of sixty men each, in successive waves. They must not be married or have children; they must not be their families' only sons. They were to take the two roads Massoud had spotted and encircle the Taliban positions on the hill. He told them to cut the supply road and hold their positions while offering the Taliban a way to escape. The idea was not to force the Taliban to fight to the last man. The idea was just to overtake their positions with as few casualties as possible.

The commanders filed out, and Reza and Dr. Abdullah were left alone with Massoud. He was exhausted, and he lay down on his side with his coat over him and his hands folded under his cheek. He fell asleep, woke up, asked Reza a question, then fell asleep, over and over again for the next hour. Occasionally a commander would walk in, and Massoud would ask if he'd repositioned those mortars or distributed the fifty thousand rounds of ammunition to the front. At one point, he asked Reza which country he liked best of all the ones he'd worked in.

"Afghanistan, of course," Reza said.

"Have you been to Africa?"

"Yes."

"Have you been to Rwanda?"

"Yes."

"What happened there? Why those massacres?"

Reza tried to explain. After a few minutes, Massoud sat up. "A few years ago in Kabul, I thought the war was finished, and I started building a home in Panjshir," he said. "A room for my children, a room for me and my wife, and a big library for all my books. I've kept all my books. I've put them in boxes, hoping one day I'll be able to put them on the shelves and I'll be able to read them. But the house is still unfinished, and the books are still in their boxes. I don't know when I'll be able to read my books."

Finally Massoud bade Reza and Dr. Abdullah good night and then lay back down and went to sleep for good. Though he holds the post of vice-president in Rabbani's deposed government, Massoud is a man with few aspirations as a political leader, no apparent desire for power. Over and over he has rejected appeals from his friends and allies to take a more active role in the politics of his country. The Koran says that war is such a catastrophe it must be brought to an end as quickly as possible and by any means necessary. That, perhaps, is why Massoud has devoted himself exclusively to waging war.

I woke up at dawn. The sky was pale blue and promised a warm, clear day, which meant that the offensive was on. Reza and I ate some bread and drank tea and then went outside with the fighters. There seemed to be more of them milling around, and they were talking less than usual. They stood in tense little groups in the morning sun, waiting for Commander Massoud to emerge from his quarters.

The artillery fire started up in late morning, a dull smattering of explosions on the front line and the occasional heavy boom of a nearby tank. The plan was for Massoud's forces to attack at dusk along the ridge, drawing attention to that part of the front, and then around midnight other attacks would be launched farther south. That was where the front passed close to Taloqan. As the afternoon went by, the artillery fire became more and more regular, and then suddenly at five-fifteen a spate of radio calls came into the bunker. Massoud stood up and went outside.

It had begun. Explosions flashed continually against the Taliban positions on the ridge, and rockets started streaking back and forth across the dark valley. We could see the lights from three Taliban tanks that were making their way along the ridgeline to reinforce positions that were getting overrun. A local cameraman named Yusuf had shown me footage of seasoned mujahidin attacking a hill, and I was surprised by how calm and purposeful the process was. In his video the men moved forward at a crouch, stopping to shoot from time to time and then moving forward again until they had reached the top of the hill. They never stopped advancing, and they never went faster than a walk.

Unfortunately, I doubted that the battle I was watching was being conducted with such grace. They were just kids up there, mostly, on the hill in the dark with the land mines and the machine-gun fire and the Taliban tanks. Massoud was yelling on the radio a lot, long bursts of Dari and then short silences while whoever he was talking to tried to explain himself. Things were not going well, it seemed. Some of the commanders weren't on the front line, where they belonged, and their men had gone straight up the hill instead of circling. As a result, they had attacked through a minefield. Massoud was in a cold fury.

"I never told you to attack from below. I knew it would be mined," he told one commander in the bunker. The man's head tipped backward with the force of Massoud's words. "The plan was not to attack directly. That's why you hit the mines. You made the same mistake last time."

The commander suggested that the mistake might have been made by the fighters on the ground.

"I don't care. These are my children, your children," Massoud shot back. "When I look at these fighters, they are like lions. The real problem is the commanders. You attacked from below and lost men to land mines. For me, even if you took the position, you lost the war."

The offensive was supposed to continue all night. Reza and I ate dinner with Massoud, then packed up our truck and set out on the long drive back to Khvajeh Baha od Din. We were leaving the front for good, and on the way out of town we decided to check in at the field hospital. It was just a big canvas tent set up in a mud-walled courtyard, lit inside by kerosene lanterns that glowed softly through the fabric. We stopped the truck and walked inside, and we were just wrapping up our conversation with the doctor when an old Soviet flatbed pulled up.

It was the first truckload of wounded, the guys who had stepped on mines. They were stunned and quiet, each face blackened by the force of a land mine blast, and their eyes cast around in confusion at the sudden activity surrounding them. The medics lifted the men off the back of the truck, carried them inside, and laid them on metal cots. A soldier standing next to me clucked his disapproval when he saw the wounds. The effect of a land mine on a person is so devastating that it is almost disorienting. It takes several minutes to understand that the sack of bones and blood and shredded cloth that you're looking at used to be a man's leg. One man lost a leg at the ankle; another man lost a leg at the shin; a third lost an entire leg to the waist. This man didn't seem to be in pain, and he didn't seem to have any understanding of what had happened to him. Both would come later. "My back hurts," he kept saying. "There's something wrong with my back."

The medics worked quickly and wordlessly in the lamplight, wrapping the stumps of the legs with gauze. The wounded men would be flown out by helicopter the next day and would eventually wind up in a hospital in Tajikistan. "*This* is the war," Reza hissed over and over again as he shot photos. "*This* is what war means."

Reza had covered a lot of wars and seen plenty of this in his life, but I hadn't. I ducked out of the tent and stood in the cold darkness, leaning against a wall. Dogs were barking in the distance, and a soldier shouted into his radio that the wounded were coming in and they needed more medicine, *now.* I thought about what Reza had said, and after a while I went back inside. This is the war too, and you have to look straight at it, I told myself. You have to look straight at all of it or you have no business being here at all.

MASSOUD'S LAST CONQUEST

2002

An unnatural fluttering of the plastic over our windows woke me. It sucked in and snapped back three times, as if the whole world were out of breath, and then it lay quiet.

A gray light leaked into the room. Dogs were barking somewhere across the fields. I got up and pulled on my clothes and climbed onto the mud roof of the house we were staying in. The moon was midway in the sky, waning toward Ramadan, and the east was shot with red. A single B-52 bomber was making its way silently across the sky at 30,000 feet, laying four thin contrails out behind. It continued past me and then made a perfect arc far to the south, where the front lines were.

I couldn't hear the bombs—they were 20 miles away—but I could feel them: four distinct pressure waves in the air that bumped past me and on up the valley. A few days earlier I'd talked to a mujahid who had fought the Russians in the 1980s. He described a Russian rocket hitting the mouth of a cave he was hiding in. The explosion itself didn't touch him, he said, but the concussion had made his ears and eyes bleed for days. That was just a Russian rocket; these were 2,000-pound bombs.

The Americans had been bombing Taliban air-defense and command-and-control bunkers across the country for a month, but it was only a few days earlier that they actually started to hit fighters on the front line. We had been waiting weeks for that to happen, speculating that diplomatic pressure from Pakistan was causing the delay. When the front-line bombing finally started, however, it was ferocious. Northern Alliance troops listened in on Taliban radio frequencies and said they could hear the Taliban screaming in their bunkers as the bombs came down; they said they could hear commanders vainly trying to contact positions that had existed only minutes before. One Taliban survivor said that 3 out of 10 men on the front line were killed before the fighting even began. At one site in Kabul, 270 foreign volunteers—mainly Arabs and Pakistanis—were killed in a single bombing run. Sixty of them were never found; they just became more Afghan dust. Locals buried what was left of the other 210.

I went back down from the roof to make some coffee. A major offensive was supposed to start within days: 12,000 Northern Alliance troops would hurl themselves at the Taliban lines and battle their way toward Kabul. The Northern Alliance had been fighting the Taliban since the mid-90s, and the State Department had been looking for Osama bin Laden for even longer. What was about to happen was a rare convergence of the interests of the most powerful nation in the world and one of the poorest. And neither could get what it wanted without the other.

I'd been in a town called Jabal Saraj—well removed from the front line—for a week or more, but now it was time to move south. I was working with three other journalists; we packed our trucks with food and cameras and flak jackets and a spare generator, and started off through the ruined fields and gutted towns of the Shomali Plain. I left most of my gear in Jabal, but I had an army rucksack with a sleeping bag and a few personal things, as well as a copy of the Koran. I tried to read it whenever we were waiting around for something, but it was slow going; like much of the Bible, it seemed to be just a long series of miseries and injunctions. One of the early lines was startling, though: *This book is not to be doubted.*

Those were fittingly stern words from a God that presided over such brutal terrain, such tortured history. The Afghans have been fighting for 23 years. Well over 5 percent of the population have been killed; more than a quarter have been displaced from their homes. Infant mortality is around 25 percent.

Tomorrow, or the next day, or the day after that, we would witness something few journalists have ever seen: a massed-infantry assault through minefields on entrenched enemy positions. Almost no one fights like that anymore—except in Afghanistan. We drove down to the front that afternoon; by nightfall, word had come that Mazar-e-Sharif, the most important Taliban stronghold in the North, had fallen.

The offensive had begun.

Last April, a delegation of Afghans arrived in Paris to plead their case with the French government and the European Parliament. At their head was Ahmed Shah Massoud, the Tajik guerrilla fighter who almost single-handedly was holding together the fractious Northern Alliance against the Taliban. With him were commanders—some would call them warlords—from the Uzbek, Hazara, and Pashtun populations. Massoud had often been accused of seeking a regime dominated by Tajiks; his decision to bring along other commanders seemed to be a message to the world that the time was right for a multi-ethnic Afghan government.

These men had grown up in villages where rice was still winnowed by hand and houses were made of wattle and mud. They had fought the Russians for ten years, and the Taliban for another five. With the exception of quick trips to Pakistan or Tajikistan, none of them had ever been out of their beautiful, war-ruined country. They landed at Le Bourget, the military airport outside Paris, were greeted on the tarmac by a woman in a short skirt, and then were taken by full diplomatic escort to the Hotel Plaza Athénée.

Massoud wore his customary safari jacket and *pakul* cap and was addressed as "commandant" by the awestruck hotel staff. The rich French food didn't agree with him, and he asked the embassy to hire a cook from a local Afghan restaurant. He was lodged in a beautiful suite with 18th-century furniture and a television. Word quickly rippled through the Afghan delegation not to turn the television on, because there were "dangerous"—i.e., pornographic—channels they might stumble onto. In some interpretations of Islam, even thinking about a woman other than your wife qualifies as a sin, and one bearded commander was observed gripping his armchair and praying, eyes closed, as a young French woman walked by.

While his commanders struggled and prayed, Massoud worked. He worked 18 hours a day, five days straight, meeting with journalists, with top-level ministers, with Bernard Kouchner, the founder of Médecins sans Frontières and former head of the U.N. mission in Kosovo, and with the entire European Parliament. His message was simple: Force Pakistan to stop supporting the Taliban regime and the war will end within the year. In addition, he asked for humanitarian help with the refugees and military help for the Northern Alliance, but that was secondary. Mainly, he warned that if Pakistan was not ostracized for its support of the Taliban, Afghanistan would continue to be a haven for terrorism and extremism. Ultimately, he said, the West would pay a terrible price.

"If I could say one thing to President Bush," Massoud said at a press conference, "it would be that if he doesn't take care of what is happening in Afghanistan the problem will not only hurt the Afghan people but the American people as well."

The Taliban were riddled with informers, and Massoud regularly received information about Pakistan's actions in Afghanistan and also about al-Qaeda. The foreign volunteers—mostly Arabs, many of them al-Qaeda—were the best fighters on the Taliban side, and had often held the line when regular Taliban positions were being overrun. Although American foreign policy was still stubbornly pro-Pakistani, some counterterrorism agents in the United States had decided to go the other way. Last summer I met with a high-level American intelligence officer who told me bluntly, "Counterterrorism means getting bin Laden, and the best way to do that is to help Massoud."

He didn't go into specifics, but it was clear that Massoud was feeding intelligence to the United States in exchange for some form of aid. It was a refreshingly practical solution to the terrorism problem, and it was clearly going to change the balance of power in the region. For decades, the United States had essentially followed Pakistan's lead when it came to an Afghan policy. During the Soviet occupation, America relied on Pakistan to put $3 billion worth of weapons and support into the hands of the mujahideen. It was all funneled through the ISI, the infamous Pakistani intelligence service, and many of the weapons wound up in the hands of anti-Western fanatics.

The power vacuum that followed the 1989 Soviet withdrawal was finally filled by the Taliban, the creation of fundamentalist lunatics recruited by the ISI from the refugee camps on the Afghan border. By 1996, Pakistan had created a rogue state that exported two-thirds of the world's heroin, brutalized its citizens with harsh Islamic laws, and hosted a terrorism network dedicated to destroying the West. The C.I.A. has a word for this: blowback. The Taliban are blowback; bin Laden is blowback; September 11 is the ultimate example of blowback. Blowback is what happens when a shortsighted policy comes back to haunt America in ways that are more dangerous than the original threat.

Blowback was a military problem, strictly speaking, and it hadn't troubled the State Department unduly. American paranoia about Iran and Soviet-aligned India dictated an unwavering support of Pakistan, and the possibility of an oil-and-gas pipeline across Afghanistan from Turkmenistan had lured a consortium of Western oil companies into negotiating directly with the Taliban regime. While American counterterrorism efforts struggled to contain the threat posed by Osama bin Laden, oil interests and Pakistani intelligence were holding American policy firmly by the ear.

Very slowly, however, elements in the American government were realizing how flawed and dangerous that strategy was. Sometime during 2000, they were considering another solution: Ahmed Shah Massoud.

By all accounts, the taking of Mazar-e-Sharif on November 10 was a bloodbath. American bombs blew holes in the front line, and Arab volunteers found it necessary to threaten their Afghan brothers with execution if they tried to defect. Tajiks under Commander Atta attacked from the south, and Uzbeks under General Dostum attacked from the north and east. The Uzbeks are known among Afghans for the utter carelessness with which they regard their own lives and for their terrible cruelty toward their enemies. Some attack on horseback with RPGs over their shoulders. There were 300 Taliban dead in Mazar, and they did not all die fighting.

After Mazar fell, the Taliban front line unzipped from north to south practically by the hour: Sheberghan, Samangan, Kholm, Baghlan, Pul-e-Kumri, Bamian. Every time a town fell, one of the soldiers we were with would come running, radio in hand, to announce it. We were now staying in a mud fortress with an overgrown courtyard and a pile of live Russian bombs in the corner. It was the headquarters of the governor of Bagram, and it had gunports in the walls and no electricity and no heat and no plumbing except for a mud outhouse tucked in one corner among the weeds. For $50 a day we could all sleep on the floor in one of the rooms, and for another $10 we would be given mutton and rice every night. And from the roof we could watch the Americans bomb the front lines only a couple of miles away.

The next morning we drove the last half-mile down to the front positions with a Northern Alliance political adviser who had turned up at our compound. He squeezed into our truck and told us that Northern Alliance intelligence was passing information about Arabs along to the Americans for bombing purposes, but they were keeping the Pakistanis for themselves.

"I don't care about Arabs, I just want the Punjabis," he said, using the local word for Pakistani.

"What will you do with them?"

"I'll kill them."

It was a matter of some outrage among Afghans that foreigners were in their country fighting a civil war. There is a lot of forgive-and-forget among the Afghans themselves—Dostum had fought for the Soviets for years—but outsiders are a different story. The British lost over 15,000 soldiers and civilians when they invaded Afghanistan in the early 1840s. Their bones, according to local legend, are still bleaching on a hillside outside Kabul. The Russians lost 15,000 men during their disastrous 10-year attempt to pacify the Afghans. The Pakistanis had some 15,000 fighters with the Taliban. Seduced by the rhetoric of extremist Islam, they had come to the mountains of Afghanistan prepared to die. And die they would.

This is where the dying would happen next: Bagram air base, a military facility built with Soviet aid in the 1950s, 35 miles north of Kabul. Bagram had changed hands three times in the past few years; now the Northern Alliance held it, but the Taliban front lines were just on the far side of the tarmac. The control tower was a shattered hulk of concrete with a blast hole in the sagging roof. Trees had grown up around it from 10 years of disuse. Scattered across the airfield below were rusting fuel trucks and the carcasses of destroyed Russian fighter planes and a couple of tanks with their turrets knocked off.

A logistics officer, who was responsible for bringing food and ammunition up to the forward trenches, took us into the control tower for a look. He said that the preceding week the Taliban had shot at anyone they saw in the tower, but this week it didn't seem to be a problem. We hoped he was right. We sat in the warm sun, passing around cigarettes, and watched through binoculars the Taliban bunkers smoking from a carpet bombing earlier that morning. One of the soldiers we were with, a teenager named Ahmad, picked up a two-way radio and punched in a Taliban military frequency. He found a Taliban soldier who was trying to locate a friend of his named Rafardeh.

"Rafardeh is dead," Ahmad said.

"I'm asking for Rafardeh," the Talib said again.

"Yeah, he's dead."

Ahmad went on to gloat on the radio over the towns that had fallen in the past 24 hours: Sheberghan, Mazar, Baghlan . . .

"Well, prepare your wives," the Talib answered. "Wash them well because we're coming. It's our turn now."

Another American bomb hit on the far side of the valley, and a dark column of smoke rose and dissipated in the pale-blue sky. The concussion reached us 20 seconds later. I could see plumes of dust on the plain from trucks that were resupplying the Taliban positions. Ahmad dialed around again and found another enemy soldier to talk to.

"Here's the password," he said. *"Tahayur."*

Tahayur means "ecstatic" in Dari. Ahmad was now talking to the Talib as if he were one of them.

"Have you heard? Everything has fallen in the North. What are we going to do?"

"Don't worry, we're still holding our positions," the Talib said. "Other Afghans won't harm us in the end—we're all Afghan."

"Yes," Ahmad said with a grin, "but can you imagine what they're going to do to the Pakistanis and Arabs?"

"Well," the Talib said, "who cares about them?"

"Listen, I'm not Talib," Ahmad finally told him. "I'm mujahideen. We know your passwords."

That didn't seem to faze the Talib. He just wanted to know when the Northern Alliance was going to launch its offensive.

"Why would we launch an offensive? Then you'll kill me, and I'll kill you," Ahmad pointed out. "We're just sending those American birds instead, the ones you can't even touch."

"You know what God will do with those birds," the Talib warned before he hung up, but he didn't sound very convinced. Ahmad laughed.

"For Sheberghan," he said, passing around some raisins from his pocket. "For Mazar."

We ate the raisins and sat back in the destroyed control tower and watched the Americans bomb.

Early last year Massoud was asked by a French journalist whether his fight against the Taliban was futile, given that they controlled 80 percent of the country and enjoyed massive military and economic backing from Pakistan. Massoud smiled. He had been fighting for the freedom of his country for 23 years.

"I'll tell you what I think: life goes by whether you're happy or not," he said. "Any man who looks back on his past and feels he has been of some use need have no regrets."

Afghan politics are impenetrable even to many Afghans, and Massoud's role could, in a certain light, be construed as simply a series of military victories followed by political blunders. The worst was his withdrawal from Kabul in 1996, following years of vicious fighting among rival mujahideen factions. As acting minister of defense, he had failed to overcome the ethnic factionalism that fueled the fighting and had failed to fully control his troops. Atrocities were committed in Kabul under his watch, and—though expecting otherwise would have been holding him to an impossibly high standard in a country wracked by two decades of war—some have never forgiven him.

But even Massoud's critics acknowledged that he was a master of guerrilla war. Nine times the Soviets blasted their way into the Panjshir Valley, 50 miles north of Kabul, and nine times he drove them out. The Taliban army, with its heavy reinforcement of Pakistani volunteers and even regular Pakistani Army troops, outnumbered the Northern Alliance three to one, but they were never able to deliver a decisive blow. The closer Massoud got to defeat, it seemed, the more resourceful and dangerous he became. That was never more true than in 1999, when his forces suffered a devastating rout at Bagram air base and were driven north to the Panjshir Valley. Five hundred thousand civilians walked all night—hundreds drowning in a river during the panic—to make it into the valley ahead of the Taliban tanks.

Massoud acted fast. First he dynamited the walls of the Dalang Sang gorge at the mouth of the valley, effectively locking himself in and the Taliban out. Then he traveled from village to village, mosque to mosque, preaching resistance: "Now it is the time to be ready to die," he said. "It is better to die fighting for the freedom of your country than to just live a good life, day to day."

The last time the Taliban had penned Massoud into the Panjshir, in 1996, it took him three months to organize a successful counterattack; this time he turned his forces around almost immediately. He gathered his best fighters and marched all night, coming down out of the mountains at dawn to attack across the Shomali Plain. The Taliban were taken completely by surprise; cut off from their supply lines, they were slaughtered by the hundreds. It was their worst defeat ever at Massoud's hands, and they never managed to take back the territory they lost.

When I was in Afghanistan in the fall of 2000, I talked to a Pakistani prisoner of war who had been trained by bin Laden's network at one of the terrorist camps outside Khost. His name was Khaled, and he described Massoud bitterly as the "last wall" that was keeping al-Qaeda from spreading fundamentalist Islam throughout Afghanistan and the rest of Asia. If they lost in Afghanistan, he said, they would be forced to find another country to use as a base for their global war against the West.

Khaled spoke readily, even proudly, about his plans, as did the 20 or so other prisoners who were with him. They said that they had come from all over the Islamic world to fight Massoud, and that if they were killed it didn't matter, because others would replace them. It was a religious war, they said, and it was without end.

With men like that in his prisons, it was not surprising that Massoud had taken to warning America about terrorism. Bin Laden had already used Afghanistan as one of the bases to coordinate the 1993 World Trade Center bombing, the 1998 embassy bombings in Africa, and the 2000 attack on the U.S.S. *Cole*. Clearly, more attacks were on the way. And it was no surprise that Massoud, though a devout Muslim, was one of the targets of an incredibly sophisticated anti-Western conspiracy taking shape in Afghanistan. Over the course of last summer, yet another Taliban offensive ground to a halt on the steppes of northern Afghanistan, and, according to a top commander, Massoud had just bought dozens of refurbished tanks from the Russians to bolster his defenses. The C.I.A. was apparently working with him directly, and Pakistan had even started taking some heat internationally for its support of the Taliban. After five years, alliances had started to shift, arms had started to flow. And one man was at the center of it all.

On September 9, Massoud was at his rear base, called Khoje Bahauddin, along the Amu Dar'ya River. He was late, as usual, for an important meeting of commanders on the front line, but early in the afternoon he stopped by a government-run guesthouse to greet two journalists who had been waiting for several weeks to see him. They said they worked for an Arab television network and came recommended by the Islamic Observation Center, a nonprofit organization in London. They claimed they wanted to interview him for a documentary on Afghanistan. Without the interview, they said, their entire trip would be wasted.

The men were never searched, partly because Massoud found it disrespectful and partly because they had been vouched for. Later, it would be discovered that their Belgian passports had been forged and that the Islamic Observation Center was linked to al-Qaeda. And someone would remember that the journalists had been strangely protective of their television camera, particularly while jolting down Afghanistan's dirt roads in a hired Russian jeep. The camera, as it turned out, was packed with explosives, and the men weren't journalists but al-Qaeda operatives sent by Osama bin Laden.

Massoud apologized for their long wait and settled himself into a chair in one of the guesthouse rooms. Seated around him were Masood Khalili, the Northern Alliance ambassador to India; a translator named Asem Suhail; and an Afghan journalist named Fahim, who was videotaping the interview as well. Massoud's bodyguards waited outside the door, as usual, out of respect for the journalists. While one of the Arabs busied himself setting up the television camera and tripod, the other gave Massoud a quick rundown of the questions he was going to ask. The first two were vague questions about the war, but the third one caught Massoud's attention: "What will you do," the Arab asked, "with Osama if you get him?"

Massoud tilted his head back and laughed. It was the last thing he ever did.

Word came late on the night of November 11 that the offensive was on. We were back at the governor's house in Bagram, sitting on the floor eating mutton, when a soldier came in and gave us the news. The Americans would resume bombing at dawn and then the Northern Alliance would move forward in successive waves of 2,500 men. The front line stretched 15 or 20 miles across the Shomali Plain, and at strategic positions alliance troops had been moved forward in preparation. There were more than 12,000 men on this side waiting to attack and perhaps twice that number on the Taliban side waiting to defend. It was possibly the largest troop concentration in Afghanistan since the Russian occupation 20 years earlier.

We woke up the next morning before first light, and the windows were already shaking from bomb blasts. They were close—just a few miles to our south—and seemed to be hitting the highway. An alliance intelligence officer had told us that 4,000 fresh Pakistani recruits were stationed on the two roads that led across the Shomali Plain toward Kabul. The Americans were trying to take them out.

Several hundred Northern Alliance troops had already gathered on the highway by the time we got there. They climbed down out of their trucks and tanks and lined up on the war-pocked hard-top, the first rays of sunlight stabbing their dark faces as they spread their hands open to the sky. First the commander intoned a long prayer and then one of the soldiers sang a very beautiful song that twisted and dissipated into the still air and left a terrible silence behind it until a roar broke out among the troops, and they clambered back into their trucks.

The convoy lurched forward, and we followed in our pickup, completely engulfed in a cloud of yellow dust that eliminated any understanding of where we were or where we were going. All across the valley, Northern Alliance troops were moving forward to their front-line positions. After a half-mile the trucks stopped, and we jumped down with the soldiers behind a mud bunker at a position called Du-Saraka as the tanks continued forward. The Americans were still bombing, and a half-dozen or so journalists were standing around, inattentive and expectant, when the first shell came in and put us all in the dirt. It was an 82-mm. mortar, and another followed a few minutes later. Finally a third hit right where all the journalists had been clustered moments before, scattering us like quail.

Shaken and dirt-streaked, we regrouped back at the highway. The Northern Alliance had gotten the Taliban positions in their range, and we could hear their rockets rip through the air as they crossed over our heads. Great gouts of smoke bloomed silently on a distant hilltop, followed by flat thuds as the concussions reached us 10 seconds later. It was mid-morning by now, and beyond Du-Saraka an enraged smattering of small-arms fire started, intensifying and subsiding and intensifying again before shifting off to the south. The alliance tanks were moving forward through no-man's-land, firing as they went, and the soldiers were following in their tracks to avoid the land mines. The Taliban lines were not holding. The alliance was breaking through, and the battle was rolling as they moved forward.

Around noon there was a lull in the fighting while Northern Alliance troops advanced to occupy positions abandoned by the Taliban. We moved forward as well, arriving at a mud fortress where troops had gathered for the next wave of attacks. Some wore ill-fitting combat fatigues bought from Iran, but most were local fighters in old wool jackets, baggy shalwar kameez, knit wool caps, and loafers or old sneakers broken down at the heel. All were strapped up with ammo belts and combat pouches and great loops of .30-caliber ammunition for the terrible, belt-fed PKAs they carried so casually over their shoulders.

The fighters squatted in the dust, feeding rounds into ammo clips and nibbling on pistachios. Some begged water off us because they hadn't drunk all day, and others begged cigarettes or food. They were young and giggled like schoolgirls when we talked to them, but when a nearby rocket launcher misfired they turned into men again and rushed over to load the gunner's limp and bleeding body into our pickup truck. Our driver took the man rearward to a field hospital, and we sat in the dust and waited like everyone else for the order to attack.

Rumor had it that the attack would come at two. At one o'clock, word went through the compound that three journalists had been shot off an armored personnel carrier (A.P.C.) up North during a Taliban ambush near Taloqan the night before. At 1:30, an Associated Press journalist came wobbling back into our camp with a bullet hole in the middle of his back, the bullet having been stopped by the steel plate of his flak jacket. Two o'clock came and went with just a ripple of agitation passing through the troops. We watched a Taliban missile float gracefully up into the air and then fall back hopelessly short of the American jet it had been trying to hit.

The jet rose and disappeared into the sunlight and then glinted back into existence as it dove. It had long since pulled out and banked for home when we heard the howl of four American bombs rushing downward toward us. We ducked—an utterly pointless act—and then a wall of smoke rose a half-mile away along the Taliban lines, followed almost immediately by four separate concussions that moved through us like little earthquakes.

The orders came in just minutes later; a commander held his radio up so that his troops could hear. One soldier started praying. This was it. Kabul lay just 20 miles to the south; this was the culmination of five years of war.

Massoud's funeral took place seven days after his death. His coffin was covered in flowers and transported atop an A.P.C. that moved slowly, like a ship through a sea of screaming people. Black-white-and-green Afghan flags waved in the slack wind, and children watched mutely from the side of the road, pouring handfuls of dirt over their heads in grief. The procession stretched for miles along the sparkling Panjshir River and wound its way up a desolate hilltop outside the town of Basarak, coming to a stop in front of a huge hole in the ground.

"We have lost Massoud, but there are a thousand other Massouds who will replace him!" Yunus Qanooni, the minister of the interior, shouted into a bullhorn.

"The world did not hear the suffering of the Afghan people, but now they have started to because the same thing has happened to them," proclaimed Burhanuddin Rabbani, the aging president of the Northern Alliance.

Indeed. Two days after Massoud was assassinated, over 3,000 people died in the attack on the World Trade Center in New York City, and over 200 more died at the Pentagon and at the crash site in Pennsylvania. It was the worst act of terrorism ever in America, eclipsing Timothy McVeigh's 1995 attack on the Alfred P. Murrah building in Oklahoma City; it was perhaps the only time in history that American civilians have felt directly targeted by an act of war.

While America reeled in shock, events in Afghanistan were moving fast. Within hours of the killing of Massoud, the Taliban launched a major offensive that was clearly meant to capitalize on the confusion and panic caused by the assassination. In an attempt to keep their front lines from collapsing, the Northern Alliance denied all rumors that Massoud was dead, until September 14, when they issued a statement that he had been killed in a suicide bombing by suspected al-Qaeda operatives. Masood Khalili had survived, but lost one of his eyes and suffered shrapnel wounds to his legs. Fahim survived with just severe flash burns across the arms and neck. Asem Suhail was killed. One of the attackers died immediately; all they found of him were his legs. The other, miraculously, survived the blast and tried to flee on foot. He was gunned down by one of Massoud's bodyguards almost immediately.

The two assassins had entered Afghanistan through Pakistan and walked across the front lines somewhere north of Kabul. Their plan was to kill Massoud in late August, which would have given the Taliban several weeks to wipe out the Northern Alliance before al-Qaeda's attacks in the United States. There's no question they could have done it, and had that happened, the United States would have found itself with no allies on the ground to do its fighting for it, no bases inside Afghanistan from which to launch search-and-destroy missions. It would have been right back where the Russians were in 1980—and everyone knows how that turned out.

Fortunately, Massoud's desperately busy schedule kept him out of reach—and alive—for nearly a month. By the time his assassins finally got to him, the Taliban had run out of time; September 11 was only two days away. While the State Department indulged Pakistan's diplomatic maneuvers in the wake of the terrorist attacks, the Pentagon abandoned all such concerns and went ahead planning a joint military operation with the Northern Alliance.

Shortly before the offensive began, the State Department, at the behest of President Pervez Musharraf of Pakistan, demanded that the Northern Alliance refrain from entering Kabul. As those words were being spoken, U.S. warplanes were busy bombing a path for them through the Taliban front lines. They were two completely contradictory messages, but it was clear which one the Northern Alliance would listen to.

The tanks went through the front lines first, and we were behind the third one in dust so thick it was as if we were moving along the floor of some sickly yellow sea. We jolted between low mud walls, past destroyed houses and the dark shapes of soldiers who loomed briefly and then slid back into the murk. There was shooting up ahead, and rockets were still ripping over our heads, but the Taliban lines had completely collapsed in the first spasms of fighting. The only thing that concerned us—and it didn't seem to bother the alliance troops much—was the danger of attacking too fast and getting cut off. The Taliban could easily repeat Massoud's tactics and just retreat into the hills and strike back at night.

All across the valley, alliance soldiers and armor were moving forward toward the two roads that led to Kabul. The tanks in front of us clanked out onto the hard-top, which was pitted by years of fighting and littered with shrapnel and spent bullets, and turned south. It was a headlong advance, and we were caught up in it like flotsam. Flatbeds full of regular-army troops and A.P.C.'s with four-barrel anti-aircraft guns and old Russian tanks and Datsun pickups packed with mujahideen all hurtled southward through no-man's-land and then past the first Taliban bunkers. We were in liberated Afghanistan.

There was no resistance at all. Groups of alliance soldiers that had battled their way through the lines were moving across the flatlands at a run and spilling out onto the highway in groups of 50, 100, 200, embracing one another, shooting their weapons into the air, throwing their arms up toward the sky. Tanks bucked to a stop, and commanders leaned down off their turrets to kiss men they recognized. A wounded man sat by himself on the side of the road, ignored in the jubilation. It was a complete rout, and the alliance wasn't going to stop until it was in Kabul.

The convoy roared southward, stopping only to absorb more troops coming in off the plain. Fires were burning on the hillsides, and Katyusha rockets hissed overhead, leaving beautiful red streaks across the sky. We rocked past blown-up cars, their contents sprayed across the hard-top, past four craters—so enormous they could have been made only by American bombs—and destroyed Russian tanks left over from the last war, still askew on the road after more than a decade.

We took some incoming fire around the destroyed town of Qara-Bagh, and the convoy accordioned to a stop, fighters spilling out of their trucks to shoot back into the darkness. It was night now, and the headlights of the tanks projected the silhouettes of men onto the dust-choked air like an old black-and-white movie. Three Taliban were dragged out of a bunker, dirty and terrified, and pushed along through the crowd toward the side of the road. One was an old man, a Turk wounded in the chest, who claimed he was a cook. A young alliance soldier cocked his gun and started to haul him off the road but was stopped by Reza, the photographer I was working with. Reza told the soldier in Dari that he had known Massoud during the 80s, when they were fighting the Russians, and that Massoud had absolutely forbidden the mistreatment of prisoners.

"I have all your photographs," Reza warned. "Respect the memory of Massoud, or I will report you all."

The Turk was put in a car with a dead alliance soldier and driven north. There was heavy fighting up ahead, and we decided it was too dangerous to continue. The journalists in Taloqan had died in a night ambush, and we wanted to be sure that didn't happen to us. Soldiers started to build twig fires along the side of the road to boil water for tea, the orange tips of their cigarettes waving in the dark. Five years of terrorism and repression had just been broken along this destroyed stretch of road.

Five dead Taliban waited for us in the middle of the highway the next morning, probably dragged out of their car and shot by alliance soldiers just hours earlier. One of them—apparently the commander—was middle-aged and fat, and lay on his back with his head thrown back. The others looked to be in their 20s and lay around him in strange contortions. Their eyes were open and they looked straight up at the sky.

There were plenty of stories of reprisal killings, but given the hatreds that had developed over two decades of war, the crimes were minimal—hundreds of dead, maybe, but not thousands. For the most part, the local Taliban were spared, and the foreigners either died fighting or were killed as they surrendered. There were also stories—plenty of them—of alliance soldiers intervening to save Taliban who were being lynched by mobs in Kabul. Those incidents, however, tended not to be as eagerly reported by the press.

We arrived outside Kabul early the next morning. Dozens of alliance tanks and several thousand fighters had stopped on the last hilltop before town, on the orders of their officers, while special units went forward to secure the city. The residents were so terrified of a power vacuum that they sent a delegation of elders up the road to beg their liberators to enter Kabul. The alliance was too worried about international criticism to respond immediately, but by noon alliance tanks were rolling through the streets. For all the U.S. State Department's hand-wringing about the Northern Alliance, it was clear that the people of Kabul wanted them in the city as soon as possible.

We left our truck where the tanks had blocked the road, and walked down off the hill, Kabul stretching out before us. Thousands of city residents had walked up the highway to greet the Northern Alliance, and we dodged through them to shouts of "America!" and "Massoud!" One kid pedaled past on a bicycle, playing a harmonica. A man showed off a curved dagger he'd taken off an Arab. The Taliban military headquarters on the outskirts had been pancaked by American bombs, and a dead Arab lay in front of it. At a street market, people were dancing in front of an old stereo that was blasting Indian rock through blown-out speakers. They were a people who had been let out of jail, and they wandered the streets with the same stunned disbelief.

And the dead: five Arabs scattered across an intersection after their truck was hit by an American rocket, eight more in a city park after a shoot-out with alliance soldiers. (Apparently the Arabs had awoken in their bunker that morning, unaware that anything was wrong, and had wandered out into the city to find their army gone.) Once the Taliban had started to unravel, they went fast. Their strongest positions were around Mazar, where al-Qaeda's 55th Brigade was stationed, and after that city fell, people knew it was only a matter of time. Many of the captured frontline fighters had been in Afghanistan only a week or two, which led to suspicions that they had been placed there to slow the alliance troops while more senior Taliban made their escape.

In Kabul, the Taliban had been in full flight by the time dark had fallen the night before. For days people had been hiding their cars, worried that desperate Taliban soldiers would steal them in order to escape, and around six o'clock that night they noticed a lot of Taliban in the streets, loading their belongings into pickup trucks. The Taliban drove out of the city in convoys headed south. Some stopped to clean out the money changers' market; others stopped to loot the national bank. Around midnight, the cook at the infamous Pul-i-Charkhi prison knocked the locks off the main gate and freed the inmates. Thousands of them, including captured alliance fighters, poured out of their cells and scattered in every direction across the dark plain, ignorant of why they'd been liberated and probably not caring.

The city changed by the day, practically by the hour. Under the Taliban it had been illegal to reproduce the human face or to do anything that would distract from Islam. Now the television station was broadcasting for the first time in five years. People started to take paintings and photographs out of the closets and to dig up chess sets from their backyards. We stayed about a week and then drove north across the old front line and then past Bagram and Jabal and into the Panjshir Valley. It was going to take a couple of days to get a helicopter out, and one afternoon Reza and I climbed up to a hilltop grave that overlooked the valley. Two middle-aged men were there, and we asked them who lay buried.

"Abdi Mohammed," one man said. "He died attacking a Russian tank that was dug in right where you're standing. He was 23 years old."

Mohammed's house was nearby. He had joined the mujahideen when the Russians took over his village, and died a few hundred yards away from the house he was born in, trying to drive them out.

A lot of things could happen now, I thought. Kabul was free, and the Taliban had been toppled, but the point of all this was to end terrorism, and that may or may not happen. There are many good reasons for doing something, though, and some don't become clear immediately. Some take decades. A short distance from Abdi Mohammed's grave was a stone bench with a sapling near it. The sapling was about five feet tall, and I asked the man why it was there.

"Well, this is a good place to sit and think," the man said. "But there was no shade here. So I planted a tree."

"When do you think it's going to be big enough to sit under?"

"Probably not for 50 years," he said. He must have seen me frown. "It's not for me, obviously," he added. "It's for the others."

Afterword

I first heard the news from my translator. Her cell phone rang in the middle of an interview. She answered, then covered the mouthpiece of her phone and said that a plane had just flown into the Pentagon. She listened again and said that another one had flown into the World Trade Center and there were ten more planes in the sky looking for targets and that the United States was at war.

We were in the town of Cahul, in southern Moldova, reporting a story on trafficking in women and forced prostitution. Moldova is, by far, the poorest country in Europe; a full quarter of the population has emigrated in search of work. No one around us knew anything about what was happening in the United States, and even the local radio station had no access to international news. The best my translator could do to confirm the story was to call her mother in Chişinău. Her mother didn't speak English, but at least she had a television. "Turn on CNN and tell me what you see," my translator said. Her mother did and immediately burst into tears. That was all we needed to know.

We continued reporting our story for the rest of that week, but it was difficult to convince ourselves that it had any relevance at all in a world so radically changed. Toppled, the World Trade Center cast a longer shadow than it ever had whole, and all journalism seemed to have become the chronicling of one day of terror and its enormous consequences. Several times, people in Moldova came up to me to shake my hand or squeeze my shoulder and say, "I'm so sorry what happened to your country," and I didn't know how to explain to them that we weren't talking about just my country. What had happened to New York had happened to the whole world, and the consequences would reverberate not just through Afghanistan but everywhere else as well—perhaps even in Moldova. We were all in this together now.

The impact of September 11 was a confirmation of something that journalists had been saying for years: The whole world is interconnected; instability in one country can eventually lead to instability everywhere. We're all locals now, in a sense. If one country is at war, every other country is at risk; if people in one country die, people in another country might be at risk. Journalists had been reporting on the chaos and suffering in Afghanistan for 23 years, and few people in the United States had listened. It was that very chaos, however, that allowed terrorism to take hold in Afghanistan, and September 11 made American readers desperate for every scrap of information about that country they could get their hands on. Books on Islam became bestsellers; magazine editors sent reporters all over the Arab world; Americans found themselves arguing about Middle Eastern politics or the severity of winter in Afghanistan. To have readers so desperately concerned about the far-off and the arcane was a journalist's fantasy of sorts—except that it came at such a terrible price.

It was not only readers who were changed, however; journalists themselves were transformed. One of the things that makes foreign reporting possible, psychologically, is the fact that you can always go home. Knowing that makes all the difference when you're shivering in a trench and haven't eaten in two days. But watching television footage of the World Trade Center collapse, I realized that was no longer true. On September 11, journalists who had covered dozens of wars—who had been shot at, who had been wounded, who had witnessed massacres and riots and starvation and disease—were reduced to tears by what they saw on television. Instead of going to war, war had come to them, and it was a different game entirely. There was no sanctuary any longer.

A good friend of mine once made the point that she wasn't a war reporter; rather, she was a human rights reporter, because the crushing of human rights is one of the things that makes war newsworthy. If modern war were fought by soldiers out in the desert or the jungle, with no involvement of the civilian population, then perhaps it would be less horrible. But today's wars kill far more civilians than combatants, and as a result, foreign wars are the most important thing that can be put on the front page of a newspaper. Similarly, we may one day understand that the "war on terrorism," as the American reaction to September 11 has come to be called, should actually be called a war on oppression. One of the striking things about the countries where al-Qaeda was born and flourished—Egypt, Saudi Arabia, Pakistan— is that they all have violently repressive regimes. Not only are they repressive, they are longtime allies of the United States. Far from protecting us, the American government's tendency to ignore human rights abuses by our friends may one day be shown to have encouraged terror and radicalism to flourish in the very countries we supported.

Rather than missiles and bombs and allegiances with repressive regimes, the ultimate military defense for America—for any country— may well be a radical and uncompromising insistence on democracy and human rights around the world. And foreign correspondence, post–September 11, may well become the best means of illustrating that one basic principle. After the Iranian revolution, journalists pointed out that American support for the Shah—a brutal dictator who spent tens of millions on himself while his country suffered— made the rise of fundamentalism and anti-Americanism in that country almost inevitable. America's fear of post-revolution Iran, in turn, resulted in its unreserved support for both Saddam Hussein's murderous regime in Iraq and Zia ul-Haq's regime in Pakistan. Under

both, virulent anti-Western sentiment flourished, and both countries
have been directly linked to al-Qaeda operations against the United
States.

Americans may well ask themselves, then, what role their foreign
policy has played in inadvertently fostering Muslim extremism
throughout the world. Journalism can encourage—can force—
Americans to ask those difficult questions. Looking back over the
problems in Afghanistan, some say they can see September 11 taking
shape even ten or twenty years ago. Looking forward, we may already
be able to discern the next generation of unthinkable headlines in the
troubles of Iraq, Indonesia, and the Middle East.

The question isn't whether we're capable of protecting ourselves.
I'm sure we are. The real question—the one I'm almost afraid to ask—
is whether September 11 was bad enough to get us to.

Acknowledgments

My research into wildfire began when Frank Carroll—then with the Boise National Forest—ignored the fact that I had no press credentials whatsoever, and wrote me a journalist pass that allowed me access to the Flicker Creek fire. My first thanks must go to him. I then embarked on a career of foreign reporting that started with a completely ill-prepared trip to Sarajevo in 1993. I am deeply indebted to Harald Doornbos—a first-rate reporter and now a good friend—who gave me a place to stay, showed me the ropes, and generally kept me out of trouble while I was there. After that I had the luxury of going to foreign countries on assignment for magazines. I would like to thank John Atwood and John Rasmus, formerly at *Men's Journal;* Hampton Sides at *Outside;* Ned Zeman, Doug Stumpf, and Graydon Carter at *Vanity Fair;* and Steve Byers at *Adventure.* I am particularly indebted to Steve Byers, who—while at *Men's Journal*—bought my first national magazine piece, and years later sent me to Afghanistan to profile Ahmed Shah Massoud. I would also like to single out Graydon Carter, the editor in chief at *Vanity Fair,* for publishing what—to many of his readers—were extremely upsetting articles about war. I'm sure it was not an easy thing for an editor to do, and I deeply appreciate his confidence in my work. I would also like to thank Starling Lawrence and Drake Bennett at W. W. Norton, as well as the many people there and at HarperCollins, for their great work and long-standing support. Photographer Teun Voeten accompanied me on many of my magazine assignments, and I would like to thank him for his companionship and his great work. And working with photographer Reza Deghati in Afghanistan was an experience that profoundly changed me, both as a person and as a journalist. And the book wouldn't exist at all without the great advice and friendship of my agent, Stuart Krichevsky, as well as Paula Balzer and Shana Cohen at his office.

Then there are my family and friends, without whom I would not be the person I am, and therefore not the journalist either. That said, I would like to thank Janine DiGiovanni, Scott Anderson, John Falk, Rob Leaver, Don Beal, Khristine Hopkins, John Vaillant, Emery Vaillant, John Evans, Victoria Bruce, Shane Dubow, Amy Kimball, Jackie Ginley, and Stephen Zanichkowsky for their interest and involvement in my work. My mother and father, Ellen and Miguel Junger, and my sister, Carlotta, have also been tremendously supportive. I am sorry for the times that I have worried them while I was on assignment. I have dedicated this book to Ellis Settle, the uncle of my oldest friend, a man of incredible wisdom whose unshakable respect for the working men and women of this world I have tried to emulate my entire life. Without the great privilege of knowing him and his wife, Joanna, I would very possibly not be a journalist at all.

Finally, I would like to point out that the profession of foreign reporting wouldn't even exist without the help of local people whose names never appear in articles, whose faces never appear on television screens. I would like to reserve my final thanks and admiration for the translators, fixers, drivers, guides, and sources who have helped me in the various countries I have worked. As locals, they take risks that no one with a foreign passport even has to contemplate. These people are too many to designate by name, but the crucial role they play in journalism should always be remembered.

(((LISTEN TO)))

FIRE

READ BY
SEBASTIAN JUNGER,
WITH
KEVIN CONWAY

$34.95 ($52.50 Can.)
8 Hours • 6 Cassettes
ISBN 0-06-000062-7
UNABRIDGED

$39.95 ($59.95 Can.)
8 Hours • 7 CDs
ISBN 0-06-000061-9
UNABRIDGED

$18.95 ($28.50 Can.)
3 Hours • 2 Cassettes
ISBN 0-06-000059-7
ABRIDGED

"[Junger has]
instinctive storytelling
skills and [an] ability
to convey a palpable
sense of the craft and
professional techniques
of a vocation."
—*New York Times*

Available wherever books are sold,
or call 1-800-331-3761 to order.

♦♦ HarperAudio
An Imprint of HarperCollins*Publishers*
www.harperaudio.com

Perennial

Books by Sebastian Junger:

THE PERFECT STORM
A True Story of Men Against the Sea
ISBN 0-06-097747-7 (paperback)

It was "the perfect storm"—a tempest that may happen only once in a century—a nor'easter created by so rare a combination of factors that it could not possibly have been worse. Creating waves ten stories high and winds of 120 miles an hour, the storm whipped the sea to levels few people on Earth have ever witnessed. Those few include the six-man crew of the *Andrea Gail*, a commercial fishing boat tragically headed towards the storm's hellish center.

"Thrilling.... Even if you have never been to sea, Junger's account will put the frighteners on you." —*New York Times Book Review*

"[A] white-knuckle chronicle . . . [A] true adventure story." —*Newsweek*

FIRE
ISBN 0-06-008861-3 (paperback)

Written with a keen eye for detail and resonating prose, Junger's new collection of nonfiction essays will take you places you wouldn't dream of going on your own.

"Where Junger excels is in his unflinching exploration of man's viciousness to man. He has an attraction for what bureaucrats and military commanders like to call 'hot zones,' places where to even set foot is to jeopardize your health."—*Newsday*

"From wildfire lightning in the American West to whale hunting in the Caribbean, the stories in *FIRE* cut straight to the heart of whatever Junger is reporting." —*New York Times*

Available wherever books are sold, or call 1-800-331-3761 to order.